T0192470

# Deployment of Rare Earth Materials in Microware Devices, RF Transmitters, and Laser Systems

# Deployment of Rare Earth Materials in Microware Devices, RF Transmitters, and Laser Systems

A. R. Jha

CRC Press
Taylor & Francis Group
Boca Raton London New York

CRC Press is an imprint of the
Taylor & Francis Group, an **informa** business
AN AUERBACH BOOK

CRC Press
Taylor & Francis Group
6000 Broken Sound Parkway NW, Suite 300
Boca Raton, FL 33487-2742

First issued in paperback 2022

© 2019 by Taylor & Francis Group, LLC
CRC Press is an imprint of Taylor & Francis Group, an Informa business

No claim to original U.S. Government works

ISBN 13: 978-1-138-05774-6 (hbk)
ISBN 13: 978-1-03-240140-9 (pbk)

DOI: 10.1201/9781315164670

Publisher's Note
The publisher has gone to great lengths to ensure the quality of this reprint but points out that some imperfections in the original copies may be apparent.

**Visit the Taylor & Francis Web site at**
**http://www.taylorandfrancis.com**

**and the CRC Press Web site at**
**http://www.crcpress.com**

I dedicate this book to my daughter, Sarita Jha for encouraging me to continue to write and for assisting with editing.

# Contents

# Foreword

This book describes the unique features of rare earth materials. It comes at a time when the future industrial development of nations and global monetary are strictly dependent on the unique applications of rare earth materials such as uranium, plutonium, cobalt, platinum, and others. In addition, supplies of oil, gas, and electricity are available at reasonable prices. Studies performed by the author seem to indicate that gas prices are fairly stable but oil prices are unpredictable due to disagreements between the oil-producing countries known as OPEC. Energy produced by wind turbine is strictly dependent on wind velocity and direction. Thus, a constant supply of electrical energy might not be possible. It should be mentioned that a large amount of electrical energy is available from steam power plants and nuclear power plants. Electricity produced by steam power plants is cheaper compared to that from nuclear power plants. Note maximum electricity is generated by nuclear plants using rare earth elements such as natural or enriched uranium but the installation cost and cost per Kw electricity is high. The electricity generated by uranium, the critical components of a nuclear reactor, the cooling process of reactors, and the reflector that use rare-earth materials are described in great detail for the benefit of the readers.

This book focuses on unique applications of rare earth materials (REMs), their compounds, and other important oxides. The book identifies significant improvements in the mechanical properties of thin membranes at high temperatures. The use of REM materials in power electronic ceramic compounds such as BNC (Ba $Nd_2Ti_4O_{12}$) and BIT ($Bi_4Ti_3O_{12}$) is best suited for electronic power circuit applications.

Neodymium and samarium-cobalt based permanent magnets offer outstanding magnetic properties at operating temperatures close to 300°C and are ideal for use in high-power travelling wave tube amplifiers (TWTAs), especially when operating in after-burner conditions.

Indium phosphate Gunn diodes are best suited to generate RF power exceeding 100 mW for radars and RF sensors tracking and detecting targets at long ranges. Solid state devices such as GaAs HEMT devices are widely used for amplifiers operating at 35 GHz and higher frequencies. GaAs devices are best suited for solid state lasers operating at 2- to 12-micron wavelengths. Forward looking infrared (FLIR) sensors are best suited for military applications such as detection, tracking, and identification of ground-based targets located in field of views ranging from 5 to 10 degrees, from 10 to 15 degrees, and from 15 to 35 degrees, and at distances ranging from 5 to 35 miles.

GaAs substrates are widely deployed in the design and development of RF circuit applications up to 35 GHz frequency and for RF devices operating at room temperatures. Note that these devices can operate at higher RF frequencies and with better RF performance if the superconducting cooling is used.

The author has published a book entitled *Next-Generation Batteries and Fuel Cells for Commercial, Military, and Space Applications* where a full chapter is dedicated to fuel cells. Performance data on proton exchange membrane fuel cells (PEMFCs) and direct methanol fuel cells (DMFCs) indicate that these cells offer high-power density and cell potential with minimum cost. Research studies further indicate that silicon-based DMFCs yield optimum electrical output power per square inch of electrode surface area.

This book is well organized and provides mathematical expressions to estimate critical performance parameters. The author clearly identifies the cost-effective features, reliability aspects, and safety aspects of the equipment that uses rare earth materials. The book especially provides a treatment of the underlying thermodynamics evaluation of the systems using REMs. The author identifies the laws of electrochemical kinetics and recommends that an appropriate rare earth material be deployed in nuclear reactors that develop with low neutron flux density. Other applications of REMs, namely underwater reactors for submarines with emphasis on nuclear reactor-operating guides, reactor cooling technology, and scheduled maintenance requirements are discussed. The deployment of nuclear technology in medical, commercial, and industrial applications is briefly mentioned in relation to products such as color television sets, permanent magnets for airborne RF jammers, high-capacity fuel cells, specialized medical equipment, and a host of commercial applications.

I highly recommend this book to a broad audience including undergraduate and graduate students, clinical scientists, aerospace engineers, nuclear engineers, project managers, life science scientists, space scientists, and design and development managers deeply involved in nuclear projects.

**Ramesh N. Chaubey**
MS (NYU Poly), PE (NY and CA)
Engineering Manager and Technical Advisor
for Infrared Military Satellites (Retired),
United States Air Force

# Author

**A. R. Jha**, PhD, has written 10 high-technology books and published more than 75 technical papers. He has worked for General Electric, Raytheon, and Northrop Grumman, and he has extensive and comprehensive research, development, and design experience in the fields of radar, high-power lasers, electronic warfare systems, microwaves, mm-wave antennas for various applications, nanotechnology-based sensors and devices, photonic devices, and other electronic components for commercial, military, and space applications. Dr. Jha holds a patent for mm-wave antennas in satellite communications. He earned a BS in engineering (electrical) at Aligarh Muslim University in 1954, an MS (electrical and mechanical) at Johns Hopkins University, and a PhD at Lehigh University.

# 1

# PROPERTIES AND APPLICATIONS OF RARE EARTH OXIDES, ALLOYS, AND COMPOUNDS

## 1.0 Introduction

This chapter deals exclusively with the properties and applications of rare earth oxides, alloys, and compounds, which might have potential uses in commercial, industrial, and military applications. Performance improvement due to deployment of these rare earth oxides, alloys, and compounds will be given serious consideration.

### 1.0.1 Mining and Processing of Rare Earth Materials

Research studies performed by the author seem to indicate that oxides of rare earth elements (REEs) are found mostly in raw conditions under the earth layers and may require processing based on grain size, prioritization, growth, and critical characteristics. It is interesting to note that prioritization is strictly based on the refining technique deployed, with particular emphasis on quality control techniques. Studies further indicate that the processing of rare earth oxides could involve several complex steps, which might be time consuming and cause additional costs. Research studies further indicate that the procurement cost per pound of oxide of a heavy rare earth element (HREE) is much higher than of a light rare earth element (LREE). In brief, the costs for LREE oxides per pound are much lower for the oxides of lanthanum (La), cerium (Ce), praseodymium (Pr), or neodymium (Nd).

*1.0.2 Mining and Processing of Rare Earth Oxides*

The availability of REEs, their oxides, and alloys is strictly dependent on the mining and processing of rare earth materials. To obtain specific REEs and their oxides, extensive efforts are required to extract the rare earth minerals deposits in certain geographical locations in selected countries. Mining companies must obtain clearance from the host countries and identify the specific locations where they plan to undertake mining and processing activities. The mining and processing efforts required to extract rare earth materials can be described as follows:

- Mining of rare earth ores can be obtained from known mineral deposits at specific locations.
- Extraction of REEs and their oxides from their known locations: finding the known location requires comprehensive metrological and mining surveys to find the desired location with minimum cost and efforts.
- Separation of rare earth ores into individual rare earth oxides is very cumbersome and costly and requires extensive efforts to satisfy specific application requirements.
- Conversion of the rare earth alloys into compounds such as samarium–cobalt magnets for specific commercial and defense applications requires further efforts, which must be accomplished with minimum cost without any compromise in quality control specifications particularly for defense-related applications.
- Deployment of rare earth alloys such as neodymium-iron-boron magnets for high-power commercial and industrial applications requires cost-effective quality control techniques to maintain uniform magnetic properties within the magnetic core medium. It is important to point out that such magnets are widely used in motors and generators, which are the most critical components of hybrid electric and electric vehicles. Furthermore, significant reduction in weight and size of the motors and generators is realized by using rare earth–based magnet technology. It is critically important to mention that such magnets offer high reliability under high vibrations and temperature environments. Laboratory tests reveal these magnets retain their magnetic characteristics close to a temperature as high as 200°C.

- As stated earlier, important quality control procedures and design specifications such as visual inspection, verification of critical performance, and reliability of data must be verified for these magnets for military components such as travelling wave amplifiers (TWAs), where the collector and anode temperatures can approach 250°C or higher.

The following popular rare earth oxides are widely used in commercial, industrial, and defense applications, where consistent operation and high reliability under rough mechanical and thermal environments are the principal design requirements. REEs are used in high-performance steel and other alloy compounds best suited for missiles, jet engines, tanks, and other military vehicles. The author wishes to highlight the industrial revolution period of 1891 to 1930 in which REEs were identified for specific industrial applications. Based on comprehensive scientific research during the period from 1930 to 1955, the properties of rare earth materials and their oxides were established. In addition, specific applications of certain REEs, alloys, compounds, and oxides were identified with particular emphasis on safety while dealing with them.

*1.0.3 Rare Earth Oxides*

Extreme care must be exercised in handling REEs and their oxides, alloys, and compounds. The following are the most popular rare earth oxides, which have a variety of commercial, industrial, scientific, and military applications:

- Thulium oxide ($Tm_2O_3$)
- Cerium oxide ($Ce_2O_3$)
- Ytterbium oxide ($Yb_2O_3$)
- Cerium oxide ($Ce_2O_3$)
- Dysprosium oxide ($Dy_2O_3$)
- Europium oxide ($Eu_2O_3$)
- Gadolinium oxide ($Gd_2O_3$)
- Holmium oxide ($Ho_2O_3$)

Research studies on REEs and their oxides performed by the author indicate that serious health problems could occur if extreme care is

**Table 1.1** Half-Lives of Certain Isotopes Used in the Medical Field

| ISOTOPE | HALF-LIFE |
|---|---|
| $Ce^{144}$ | 292 days |
| $Pr^{144}$ | 18 min |
| $La^{140}$ | 40 hrs |
| $Eu^{156}$ | 15.6 days |
| $To^{232}$ | $1.4 \times 10^{10}$ years |
| $Pu^{239}$ | $2.42 \times 10^4$ years |
| $U^{233}$ | $1.62 \times 10^5$ years |

not exercised in dealing with REEs and their oxides. There is more radiation danger from REEs compared to their oxides. Since the medical field deploys various rare earth isotopes for diagnosis and treatment of diseases, medical technicians must handle these isotopes with great care and precautions. Note that the half-lives of the rare earth isotopes determine the level of radiation danger. For the benefit of the readers the author has summarized the half-lives of certain isotopes in Table 1.1.

### 1.0.4 Estimated Worldwide Production of Rare Earth Oxides from 1950 to 2000

As stated earlier, rare earth experts and scientists were very busy with the systematic discovery of chemical and physical properties of rare earth oxides, identifying possible commercial and industrial applications and their use in medical research activities. Mass-scale medical research programs were initiated in the United States, England, Germany, France, and other European countries. Because of the high demand for rare earth oxides, global production for these oxides was given serious consideration during the period ranging from 1960 to 2000. The estimated worldwide production of rare earth oxides can be seen in Figure 1.1. This figure shows that China produced the highest amount of rare earth oxides over this period. The Chinese anticipated an increased demand for rare earth oxides and thus were able to dictate the price per pound of oxide. In addition, they set up the supply quota and price per pound particularly for Japan, because Japan was buying this product in abundance.

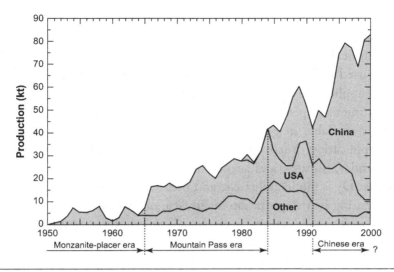

**Figure 1.1**   Global production of rare earth oxides from 1950 to 2000 (After Haxel et al., 2002).

*1.0.5 Applications of Various Rare Earth Oxides*

Comprehensive research studies performed by the author on the applications of rare earth oxides seem to indicate that these oxides are widely used in commercial, industrial, and scientific applications. Due to the abundance of dysprosium (Dy) oxide and its seven isotopes in the Earth's crust, microwave scientists and engineers decided to examine various applications for high-performance optical coating and thin-film technology.

*Example of Dysprosium Oxides Applications*   Initially, scientists and microwave engineers focused on the application of Dy oxides in mobile phones, smart phones, and computer tablets. Later, microwave engineers decided to deploy Dy oxide in the development of neodymium-iron-boron permanent magnets, dysprosium-iron-garnet optical crystals, and a host of electro-optical devices to enhance the mechanical integrity of the devices under harsh mechanical and thermal environments. This oxide is specially used in ceramic compounds, where reliable operation under harsh mechanical conditions and a wide temperature range is the principal design requirement.

It is critically important to point out that due to its emitting capability in the 470 to 500 nm region and 570 to 600 nm infrared spectral regions, this oxide is preferred for advanced optical formulations.

Furthermore, scientists found out that its high electron beam characterization is most suited for clinical therapy, tumor targeting, and other clinical applications. Due to its high susceptibility, this oxide is widely used in Dy-based compounds, which are best suited for data storage applications such as high-density compact discs.

It is interesting to mention that Dy comes in various forms. Its oxide is available in powder form or dense pellet form depending upon application. Furthermore, this oxide is available in forms such as chlorides, nitrates, and acetates. It is critically important to mention that Dy is moderately toxic and therefore extreme care should be taken during it transportation by air, sea, or road. Its room-temperature electrical conductivity is close to 57 microohm-cm and its thermal conductivity is roughly 10.7 W/meter-K. The author will specify the isotopes of selected rare earth oxides or elements that are widely used in commercial and industrial applications, such as Dy and neodymium isotopes.

### 1.0.6 Various Application of Rare Earth Isotopes

*1.0.6.1 The Dysprosium Element and Its Isotopes*   As mentioned earlier, an abundance of Dy oxide with different atomic mass numbers is available from the earth surface. The atomic mass and abundance of Dy in percent available in the Earth's crust are summarized in Table 1.2.

*1.0.6.2 Neodymium Element and Its Isotopes*   This particular element and its oxide were first discovered in 1985. It is the most abundant rare earth material in the Earth's crust after cerium and lanthanum. It is available as a metal, oxide, or compound with purity ranging from 99

**Table 1.2**   Atomic Mass and Abundance of Dysprosium Isotopes

| ISOTOPE | ATOMIC MASS | % AGE OF ISOTOPE PRESENT IN EARTH |
|---------|-------------|-----------------------------------|
| Dy-156  | 155.82      | 0.16                              |
| Dy-158  | 157.82      | 0.10                              |
| Dy-160  | 158.83      | 2.34                              |
| Dy-161  | 160.93      | 18.90                             |
| Dy-162  | 161.93      | 25.50                             |
| Dy-163  | 162.93      | 26.90                             |
| Dy-164  | 163.93      | 28.12                             |

to 99.99%. The metal is available in the form of foil, rod, or spurring target. Its compounds are available in sub-micron form or nanopowder. Primary applications of this metal include infrared lasers operating efficiently in the 2-micron spectral region, glass coloring and tinting, protecting goggles during welding operations, cathode-ray tube displays, and dielectric coatings. Its most attractive application is found in neodymium-iron-boron permanent magnets, which are widely used in electric motors and generators for hybrid electric and electric vehicles. Its potential application has been observed in the multilayer capacitors best suited for power electronic application. Other potential applications include industrial chemicals and metrological engineering fields. This particular element is widely used in the design and development of neodymium-yttrium-aluminum-garnet (Nd:YAG) lasers, which are employed for glaucoma diagnosis, upgraded multiple wavelength laser, and long-pulse Nd:YAG lasers for curing cysts. The electrical conductivity of this element is around 64 microohm-cm, while its room-temperature thermal conductivity is 16.5 W/meter-K. This element has six isotopes, which are described in Table 1.3.

**Table 1.3**  Atomic Mass and Neodymium Isotopes

| ISOTOPE | ATOMIC MASS | ABUNDANCE IN EARTH'S CRUST (%) |
|---------|-------------|-------------------------------|
| Nd-142 | 141.91 | 27.13 |
| Nd-143 | 142.91 | 12.19 |
| Nd-144 | 143.91 | 23.90 |
| Nd-146 | 145.91 | 17.19 |
| Nd-148 | 147.92 | 5.76 |
| Nd-150 | 149.92 | 5.64 |

NOTE: The author wishes to remind the readers that radioactive isotopes that include polonium (84), radium (88), thorium (90), and uranium (92) are different from the naturally occurring unstable radio-active isotopes. Furthermore, elements such as thallium (81), lead (82), and bismuth (83) are called unstable radioactive. It must be remembered that the unstable radioactive isotopes undergo spontaneous change i. e., radioactive disintegration or radioactive decay at a definite rate. However, scientists consider the rate of radioactive decay by means of half-life. Essentially, the half-life is defined as the time required to decay the radioactivity level to half its initial value.

*1.0.7 Rare Earth Alloys*

Samarium–cobalt and neodymium-iron-boron are classical examples of rare earth alloy compounds. Rare earth compounds are widely used in the design and development of high-strength magnets for microwave transmitters and travelling wave tube amplifiers (TWTAs). Note that these rare earth magnets are best suited for operations where high magnetic properties under ultra-high operating temperatures are the principal design requirements. There are other examples of rare earth compounds which are best suited for microwave filters with low insertion loss and high stop band rejection characteristics. Technical papers and published reports reveal that ferrite compound materials have been developed for the design and development of high-power magnets capable of generating strong and uniform magnetic fields for microwave transmitters and TWTAs. Such magnets have demonstrated remarkable reliability in rough mechanical environments and at elevated temperatures as high as 300°C. Furthermore, these high-temperature microwave magnets have shown no failures despite operating for long periods of time at temperatures close to 275°C.

*1.0.8 Rare Earth Elements*

After describing the properties and applications of rare earth oxide, alloys, and compounds, the author will focus on the properties and applications of REEs. It is important to stress that pure or heavy rare earth materials or elements are generally used for specific industrial, military, and scientific applications. For example, uranium, thorium, and plutonium are specifically used as reactor cores. These core materials are used especially for electrical power generation. Other rare earth materials such as beryllium, lead, and carbon are deployed to capture gamma-ray energies of reactor materials. Note that some selected REEs are of special interest in the reactor field. It is critically important to mention that hydrogen and beryllium are most effective in capturing gamma radiation emitted as a single photon per neutron capture. Now the author wishes to focus on the important properties and applications of various REEs. More detailed information will be provided on group elements, LREEs, and HREEs, their atomic numbers, atomic weights, and non-metals. The specific parameters and

**Figure 1.2**  Periodic table of rare earth elements showing critical parameters.

Periodic table (GROUP columns): categories indicated across the table — LIGHT METALS, HEAVY METALS, NON METALS, INERT GAS; sub-labels BRITTLE, DUCTILE, LOW-MELTING.

| PER100 ↓ | 1A | 2A | 3B | 4B | 5B | 6B | 7B | 8 | 8 | 8 | 1B | 2B | 3A | 4A | 5A | 6A | 7A | 0 |
|---|---|---|---|---|---|---|---|---|---|---|---|---|---|---|---|---|---|---|
| 1 | $^{+1\ -1}$ | | | | | | | | | | | | | | | | | $^{0}$ He 2 |
| 2 | $^{+1}$ Li 3 | $^{+2}$ Be 4 | BRITTLE | | | | | | DUCTILE | LOW-MELTING | | | $^{+3}$ B 5 | $^{+4}_{+2\ -4}$ C 6 | $^{+2\ -1}_{+3\ N-2}$ N 7 $^{+4}_{+5}$ | $^{-2}$ O 8 | $^{-1}$ F 9 | $^{0}$ Ne 10 |
| 3 | $^{+1}$ Na 11 | $^{+2}$ Mg 12 | | | | | | | | | | | $^{+3}$ Al 13 | $^{+2\ -4}_{+4}$ Si 14 | $^{+3\ -3}_{+5}$ P 15 | $^{+4\ -2}_{+6}$ S 16 | $^{+1\ -9}_{+7}$ Cl 17 | $^{0}$ Ar 18 |
| 4 | $^{+1}$ K 19 | $^{+2}$ Ca 20 | $^{+3}$ Sc 21 | $^{+2}_{+4}$ Ti 22 | $^{+3}_{+5}$ V 23 | $^{+2}_{+6}$ Cr 24 | $^{+3}_{+4}$ Mn 25 $^{+2}_{+7}$ | $^{+3}$ Fe 26 $^{+2}$ | $^{+3}$ Co 27 $^{+2}$ | $^{+3}$ Ni 28 $^{+2}$ | $^{+1}$ Cu 29 $^{+2}$ | $^{+2}$ Zn 30 | $^{+3}$ Ga 31 | $^{+2}_{+4}$ Ge 32 | $^{+3}_{+5}$ As 33 | $^{+4\ -2}_{+6}$ Se 34 | $^{+1\ -1}_{+5}$ Br 35 | $^{0}$ Kr 36 |
| 5 | $^{+1}$ Rb 37 | $^{+2}$ Sr 38 | $^{+3}$ Y 39 | $^{+4}$ Zr 40 | $^{+3}_{+5}$ Nb 41 | $^{+6}$ Mo 42 | $^{+4}_{+6}$ Tc 43 $^{+7}$ | $^{+3}$ Ru 44 | $^{+3}$ Rh 45 | $^{+2}$ Pd 46 $^{+4}$ | $^{+1}$ Ag 47 | $^{+2}$ Cd 48 | $^{+3}$ In 49 | $^{+2}_{+4}$ Sn 50 | $^{+3\ -2}_{+5}$ Sb 51 | $^{+4\ -2}_{+6}$ Te 52 | $^{+1\ -1}_{+5}$ I 53 $^{+7}$ | $^{0}$ Xe 54 |
| 6 | $^{+1}$ Cs 55 | $^{+2}$ Ba 56 | ♦ 57-71 | $^{+4}$ Hf 72 | $^{+5}$ Ta 73 | $^{+6}$ W 74 | $^{+5}_{+7}$ Re 75 | $^{+4}_{+9}$ Os 76 | $^{+3}$ Ir 77 $^{+4}$ | $^{+2}$ Pt 78 $^{+4}$ | $^{+1}$ Au 79 $^{+3}$ | $^{+1}_{+2}$ Hg 80 | $^{+3}$ Tl 81 $^{+1}$ | $^{+4}$ Pb 82 $^{+2}$ | $^{+3}$ Bi 83 | $^{+2}_{+4}$ Po 84 | At 85 | $^{0}$ Rn 86 |
| 7 | $^{+1}$ Fr 87 | $^{+2}$ Ra 88 | ★ 89–103 | | | | | | | | | | | | | | | |

TRANSITION ELEMENTS {BETWEEN GROUP 24 AND 34}

| ♦ LANTHANIDES (RATE EARTHS) | $^{+3}$ La 57 | $^{+3}_{+4}$ Ce 58 | $^{+3}$ Pr 59 | $^{+3}$ Nd 60 | $^{+3}$ Pm 61 | $^{+2}_{+3}$ Sm 62 | $^{+2}_{+3}$ Eu 63 | $^{+3}$ Gd 64 | $^{+3}$ Tb 65 | $^{+3}$ Dy 66 | $^{+3+}$ Ho 67 | $^{+3}$ Er 68 | $^{+3}$ Tm 69 | $^{+2}_{+3}$ Yb 70 | $^{+3}$ Lu 71 |
|---|---|---|---|---|---|---|---|---|---|---|---|---|---|---|---|
| ★ ACTINIDES | $^{+3}$ Ac 89 | $^{+4}$ Th 90 | $^{+4}_{+5}$ Pa 91 | $^{+3}_{+4\ +6}$ U 92 | $^{+3}_{+4\ +5\ +6}$ Np 93 | $^{+3}_{+4\ +5\ +6}$ Pu 94 | $^{+3}_{+4\ +6}$ Am 95 | $^{+3}$ Cm 96 | $^{+3}_{+4}$ Bk 97 | $^{+3}$ Cf 98 | Es 99 | Fm 100 | Md 101 | No 102 | Lw 103 |

KEY TO CHART — OXIDATION NUMBERS $\{^{+2\ -4}_{+4}$ Si  $14\}$ — OXIDATION NUMBER / SYMOL / ATOMIC NUMBER

properties of various REEs are shown in Figure 1.2. Light and heavy radioactive materials, group elements, lanthanides, and actinides are also clearly identified in Figure 1.2. The author hopes that readers will get familiar with the REEs, their properties, and their applications in the commercial, industrial, and military fields. Furthermore, the author wishes to remind engineers and scientists that extreme care must be taken in handling or transmitting REEs regardless of the REEs, alloy, or compound.

## 1.1  Rare Earth Compounds and Their Applications

Comprehensive studies performed by the author on rare earth compounds reveal that some of them have demonstrated unique performance in specific scientific applications. For example, the cerium-platinum (CePt) rare earth compound is best suited for the fabrication of millimeter-wave micro-electro-mechanical system

(MEMS) series and shunt switches, which require very high isolation such as 45 dB at 40 GHz. It is interesting to point out that the room-temperature electrical conductivity for both cerium and platinum is identical – $0.013 \times 10^{10}$ mho/cm. In summary, it can be stated that this particular rare earth compound also provides optimum isolation from the series and shunt radio frequency (RF) switches using MEMS technology. Note that high isolation is of critical importance for radar and electro-countermeasure (ECM) systems.

Similarly, when REE samarium is doped with a cobalt element, an important rare earth compound called samarium–cobalt is formed. This rare earth compound is widely used in the design and development of powerful permanent magnets for RF TWTAs, electric motors, and generators. Note that these motors and generators are widely used in electric and hybrid electric motors because such motors and generators have to demonstrate highest reliability at operating temperatures close to 300°C. This material is essential in the design and development of high-power, CW TWTAs widely used in extended ECM missions. Note the operating temperatures in these TWTAs have been observed to be close to 300°C and reliability is of critical importance during these planned ECM missions.

Another example of the deployment of a rare earth compound is rare earth–doped crystals, which are widely used in the design and development of infrared lasers capable of yielding high efficiency, optical stability, and excellent beam quality. A classic example of this laser involves the use of neodymium-based (yttrium-aluminium-garnet) Nd:YAG infrared laser. Note that this particular laser was used for space optical communication equipment in the late 1970s. According to space scientists, this Nd:YAG laser system provided three functions such as space-to-space communication, space-to-ground communication and space-to-ground date transfer capability. It is important to point out that the use of high-temperature ferrite ceramic played an important role in maintaining the laser beam stability under various space temperature environments. It is essential to identify critical REEs that can be used in the formation of rare earth compounds vital for potential commercial and military applications. Such REEs are summarized in Table 1.4.

**Table 1.4** Rare Earth Elements Best Suited for Development of Rare Earth Compounds

| RARE EARTH ELEMENT | RARE EARTH COMPOUND | APPLICATION |
|---|---|---|
| Cerium (Ce) | Ce Pt[3] | Best suited for MEMS switches |
| Cadmium (Cd) | Cd S | High-energy batteries |
| Cobalt (Co) | $Sm_2O_3$ | Magnets for motors and generators |
| Lanthanum (La) | La $AlO_3$ | Low-loss superconductor substrate |
| Neodymium (Nd) | Nd:YAG | Space-based IR lasers |
| Praseodymium (Pr) | Pr:Steel | Used to produce mischmetal |
| Samarium (Sm) | $Sm_2O_3$ | High-temperature TWTAs |
| Ytterbium (Y) | $YBa_2Cu_3O_2$(YBCO) | Superconductor thin film |
| Zirconium (Zr) | Yttria-stabilized zirconium(YSZ) | Superconducting substrates |

*Ref:* A. R. Jha, Superconductor Technology, John Wiley & Sons, New York, 1998, pp. 29, 30.

### 1.1.1 Rare Earth Alloys for High Temperature, High Strength Permanent Magnets

Comprehensive research studies undertaken by the author on rare earth alloys such as neodymium–cobalt and samarium–cobalt for permanent magnets designs seem to indicate that these rare earth alloys are best suited for designing the high-strength magnets that will enable magnetic alloys to retain the initial magnetic performance at elevated temperatures ranging from 270 to 300°C. This type of magnetic performance is not available from other conventional magnetic materials. The author wishes to highlight the outstanding characteristics of these rare earth alloys. Commercial and industrial applications for neodymium-based permanent magnets can be summarized as follows:

- Lifting magnets in industrial applications
- Disc magnets
- Ceramic magnets
- Magnets for physical therapy
- Radical ring magnets
- Rubber-coated magnets
- Magnets for hematite jewelry
- Magnets for motors and generators widely used in electric and hybrid-electric vehicles, where continuous operative temperatures can hover around 270°C.

It should be noted that mechanical and thermal capabilities of samarium–cobalt magnets are superior to those for neodymium–cobalt magnets. Furthermore, it must be noted that the mechanical integrity of samarium–cobalt magnets at elevated temperature as high as 300°C is outstanding. That is why the focusing magnets for TWTAs and high-power radar transmitters prefer using samarium–cobalt magnets. These magnets have demonstrated superior magnetic performance under rough mechanical and extreme thermal environments. Major commercial and industrial applications of samarium–cobalt magnets can be summarized as follows:

- Focusing magnets for space-based TWTAs, high-power magnets radar transmitters, and dual-mode TWTAs widely deployed in ECM prefer samarium–cobalt permanent magnets. It is critically important to mention that coherent magnetic field and uniform magnetic intensity under elevated temperature and rough mechanical conditions are the principal design requirements for the radar and ECM RF sources.
- Polytetrafluoroethylene or felon magnets.
- Laser pointers widely used for scientific or commercial presentations.
- Ring magnets.
- Best suited for applications where high magnetic force and elevated operating temperatures are the principal design requirements.

Time after time, rare earth alloys or compound magnets have demonstrated superior magnetic performance and high reliability under harsh thermal and mechanical environments. Comprehensive research studies undertaken by the author reveal that no other rare earth permanent magnet can beat or meet the performance levels of the neodymium-cobalt and samarium–cobalt permanent magnets.

### 1.1.2 *Applications of Rare Earth Materials*

The author wishes to identify some selected REEs that have potential commercial, industrial, and military applications. This will require comprehensive examination of specified REEs that will be best suited

**Table 1.5**   Rare Earth Elements for Specific Commercial, Industrial, and Military Applications

| RARE EARTH ELEMENT | COMMERCIAL/INDUSTRIAL APPLICATIONS | MILITARY APPLICATIONS |
|---|---|---|
| Cerium, lanthanum, europium | glass additives | communications, lasers |
| Dysprosium, cerium | hybrid vehicles, batteries | infrared lasers |
| Europium, terbium | energy-efficient light bulbs | High-intensity light sources |
| Europium, erbium | fiber optic transmission lines | fiber optic amplifiers |
| Promethium | X-ray devices | portable X-ray equipment |
| Scandium | flood lights for stadiums | high-intensity lights |
| Samarium | permanent magnets motor/generator | TWTAs, ECM jamming sources |
| Plutonium | fuel rods for nuclear power reactors | nuclear reactor cores |
| Uranium | fuel rods for power reactors | nuclear fuel for power reactors |

for various applications. The REEs summarized in Table 1.5 seem to play critical roles in specific commercial and military applications.

### 1.1.3  Critical Primary Applications of Rare Earth Elements

Under this heading the author plans to discuss specific application details for most desirable REEs and their advantages, with particular emphasis on reliability and significant reduction in weight, size, and power.

#### 1.1.3.1 Applications of Neodymium Rare Earth Metal
It is interesting to note that neodymium, cerium, and lanthanum metals are found in abundance in the Earth's crust and are widely used in commercial and industrial applications. Neodymium-YAG (yttrium-aluminum-garnet) is widely used in the design and development of infrared lasers emitting at various wavelengths, each optimized for specific scientific and medical applications. As a metal, neodymium offers several industrial scientific applications such as dielectric coatings, multi-spectral lasers, and multilayer capacitors for possible applications in high-power electronic components and systems. It should be noted that its electrical conductivity is 64 microhm-cm at an operating temperature of 1000°C, whereas its thermal conductivity is about 16.5 W/cm-K. Also of note is that Japan held a key technology patent for neodymium-based Nd-iron-boron permanent magnets, which expired at the end of 2014. This particular permanent magnet is best suited for

multiple commercial and industrial applications. Note that some rare earth–based materials have been deployed in the design and development of commercial off-the-shelf (COTS) components, which have been included in the defense system supply chain. Such components are currently deployed in mission computers and hard drives to reduce procurement costs with no compromise to reliability and safety.

*1.1.3.2 Applications of Samarium Rare Earth Metal*   The author wishes to summarize the potential applications of rare earth samarium metal. This particular rare earth metal was first discovered in 1879. It is also available as an oxide. However, the metallic version comes in various forms such as rods, wires, pellets, or sputtering targets. Thin films are best suited for precision optical coatings and high-performance capacitance for possible application in microwave circuits. As stated earlier, samarium is widely used in the design and development of high-performance permanent magnets, which are currently deployed for electric motors and generators. These motors and generators offer considerable reduction in weight and size, which is of critical importance for the auto industry. This rare earth metal in conjunction with cobalt forms a high-performance permanent magnet which can retain all its magnetic properties even at operating temperatures as high as 300°C. It is of paramount importance to mention that samarium–cobalt permanent magnets have been tested for improved reliability and magnetic performance in radar transmitters and dual-mode TWTAs deployed in airborne ECM jamming systems, where operating temperatures can rapidly approach 300°C under "after-burner" conditions. In summary, it can be stated that samarium–cobalt permanent magnets are best suited for applications where minimum weight and size, high reliability, and elevated operating temperatures are the principal design requirements.

It can be further stated that rare earth material scientists feel that this rare earth metal can be widely used in dealing with selective separation of samples based on synergistic extraction techniques, interactions between metal ions and carbohydrates, spectroscopic research studies, and chemical evaluation and characterization of certain aluminum-based rare earth metallic compounds.

Major applications of this REE include in mobile phones, high-capacity batteries, and infrared lasers operating in several spectral regions. Each spectral bandwidth is typically a narrow spectral region not exceeding 2 to 5 percent. However, infrared lasers offer high-intensity light of uniform density, and are considered extremely useful in medical applications. It should be noted that an alloy consisting of samarium and cobalt elements forms a high-intensity permanent magnet, which is widely used in the design and development of electric motors and generators, critical components for electric and hybrid electric vehicles. For the last three decades, samarium–cobalt permanent magnets have played critical roles in the design and development of high-power single-mode and dual-mode TWTAs, which are widely used especially in airborne ECM systems, where the operating temperatures approach 300°C. This high-temperature, high-power, high-reliability TWTA performance at temperatures as high as 300°C has played a key role particularly in the "after-burner mode" of the fighter-bomber jet aircraft.

TWTAs using such magnets are widely deployed in the design of high-power radar transmitters, which made radar detection and tracking performance possible over long distances. Such radar transmitters are best suited for ballistic missile early surveillance (BMES) programs and on-the-horizon detection radars. It is essential to mention that these permanent magnets have demonstrated significantly improved electron beam focusing, higher efficiency, and minimum weight, size, and power consumption. These advantages of rare earth magnetic materials have applications for airborne radar transmitter and airborne ECM jamming equipment. It critically important to mention that conventional magnetic materials will not able to meet the sophisticated operational requirements of state-of-the-art radar transmitters and high-power TWTAs for ECM equipment.

Research studies undertaken by the author seem to indicate that aluminum-based rare earth metallic compounds are used in molecular imaging as a contrast agent and in radioisotope medical research studies for diagnosis and clinical treatment of various diseases.

### 1.1.4 Properties and Applications of Cerium Material (Ce)

This REE was first discovered by German scientist Wilhelm von Hissinger in early 1803. According to material scientists, this material is the most abundant among the rare earth materials. Cerium is available as a metal oxide or compound. It is widely used in metallurgy, glass polishing, ceramics, catalysts, and phosphors-based agents. Its most common industrial application is in the steel manufacturing process. It is considered as the most efficient glass-polishing agent. This material is widely used in manufacturing medical glassware and aerospace windows because of its excellent mechanical properties, particularly the impact resistance capability. This element plays a critical role in the manufacturing of high-quality and high-strength ceramics. It is best suited for dental compositions and it acts as a phase stabilizer in zirconium-based products. It is widely employed in the manufacturing of high-performance optical components where precision, accuracy, and quality control are the principal design requirements. This material is best suited in the manufacturing of power train components due its ability to provide high accuracy and to consistently maintain accurate gear ratio regardless of operating environments. Its oxides have potential applications in the development of high-quality optical coatings and thin-film development. Its isotopes are free from radioactive effects and are widely used in medical and clinical applications.

The author has summarized the isotopes, atomic mass, and the percentage of abundance in the earth surface for cerium. Important parameters of cerium can be seen in Table 1.6.

As stated earlier, these isotopes are non-radioactive and therefore other medical and clinical applications can be explored. Further research needs to be undertaken to explore the application of these isotopes in medical treatments and clinical diagnosis for diseases.

**Table 1.6** Abundance of Cerium Isotopes in Earth Surface and Their Atomic Mass

| CERIUM ISOTOPE | ATOMIC MASS | ABUNDANCE IN EARTH SURFACE (%) |
| --- | --- | --- |
| Ce-136 | 135.91 | 0.188 |
| Ce-137 | 136.87 | 0.105 |
| Ce-138 | 137.91 | 0.249 |
| Ce-139 | 138.90 | 0.098 |
| Ce-140 | 139.91 | 88.47 |
| Ce-141 | 140.89 | 0.102 |
| Ce-142 | 141.896 | 11.07 |

In summary, it must be mentioned that cerium plays a critical role in the design and development of a hard substrate material, which is widely deployed in fiber optic transmission lines and in the design of infrared lasers operating over narrow spectral bandwidth. It is interesting to point out that cerium fiber optic transmission lines are best suited for designing the fiber optic amplifiers. Such amplifiers have demonstrated wide spectral bandwidth, higher gain, low optical transmission loss, and stable optical performance over reasonable operating temperature range.

### 1.1.5 *Applications and Properties of Rare Earth Metal Niobium (Nb)*

This metal is most popular among the microwave scientists and ASW engineers who are deeply involved with microwave filters, cryogenic cooled filters, and other RF sensors such as SQUID sensor. It is interesting to mention that the rare earth metal Niobium (Nb) is a low-temperature superconducting (LTSC) material, while lanthanum (La) is a high-temperature superconducting material. Note that a superconducting quantum interference device (SQUID) consists of one or more Josephson junctions and the associated electronics such as detection coil or loop, SQUID sensor, and control electronic circuits. It is critically important to mention that the detection loop or coil is made of superconducting material. Note that gain and SQUID sensitivity are strictly dependent on the superconducting materials deployed for the RF loops. It is interesting to point out that both a high temperature superconductor (HTSC) and a LTSC are available for the detection loop or coil, the critical element for this device.

### 1.1.6 *Applications and Properties of Rare Earth Element Yttrium (Y)*

This REE was first discovered in 1878. The room-temperature electrical conductivity of this material is 29.4 microohm-cm and its thermal conductivity is around 34.7 W/cm-K. Its atomic mass number is 167.944 and this element has only one isotope that is designated as Y-168. Research studies undertaken by the author reveal that this element has two valence states, namely +2 and +3. The studies further indicate that its isotope possesses unique properties and has commercial and industrial potential. When alloyed with iron and garnet, this

**Table 1.7**   Sintering Temperature (Ts), Critical Temperature (Tc), and Current Density (Id) for YBCO Needed for Depositing Superconducting Thin Films of YBCO Material on Various Substrates

| SUPERCONDUCTING SUBSTRATE | TS (C) | TC (K) | CURRENT DENSITY (A/CM$^2$) AT CRYOGENIC TEMPERATURE OF | |
| --- | --- | --- | --- | --- |
| | | | 4.2 K | 77 K |
| SrTiO$_3$ | 835 | 95 | $6 \times 10^6$ | $5 \times 10^{5-}$ |
| SrTiO$_3$ | 605 | 92 | $2 \times 10^6$ | $3 \times 10^6$ |
| SrTiO$_3$ | 400 | 85 | $5 \times 10^6$ | $8 \times 10^6$ |
| MgO | 720 | 87 | $8 \times 10^5$ | $10^5$ |
| MgO | 600 | 80 | $10^6$ | $10^5$ |
| MgO | 650 | 82 | $10^6$ | $10^5$ |

element forms an yttrium-iron-garnet (YIG) hard microwave substrate, which is best suited for the design and development of superconducting microwave devices using thin-film technology.

Thin films of superconducting can be deposited only on superconducting substrates such as strontium titanates (SrTiO$_3$) or magnesium oxide to achieve improved RF performance in terms of insertion loss (IL) and sintering temperature (T and the critical temperature T$_C$). These parameter requirements must be satisfied when depositing YBCO thin films on various superconducting substrates, as illustrated in Table 1.7.

It is critically important to remember that the above-mentioned properties are subject to film-growth conditions over a wide range of sputtering temperatures and oxygen pressure. In addition, the roles of equilibrium thermodynamics, kinetics, and chemical reactivity must be carefully examined.

### 1.1.7 Properties and Applications of Rare Earth Element Ytterbium (Yb)

This rare earth metal is non-radioactive and best suited for medical applications. Its most attractive property is the X-ray diffraction property, which plays a critical role in surface area and parameter analysis. Material scientists believe that it has applications in material classification. It is considered a toxic material and, therefore, its transportation and handling in the laboratory require great care and precaution. Studies indicate that the Yb-based fiber optic laser emits at 976 nm wavelength and is best suited for applications in which superior optical performance and system reliability are the principal requirements.

Ytterbium-based lasers are best known for improved beam quality, low noise level, and optical stability. Optical scientists believe that single-frequency ytterbium-doped laser performance is outstanding. The fiber optic amplifier yields excellent performance over wide spectral regions when this laser is emitting at 960 nm wavelength. Security experts believe that this particular laser is best suited for finger-mark detection capability, photo-acoustic microscopes, and energy transfer applications. Solid-state scientists believe that erbium (Er)-ytterbium thin films could be employed in second-harmonic-generation laser using external-cavity diode laser.

### 1.1.8 Properties and Applications of Thorium Element (Th)

During the 1940s and 1950s, this element was considered an important rare earth material for atomic power plants. However, comprehensive research activities in nuclear power generation later on found out that a very small amount of uranium is needed for this application. It should be noted that this element was first discovered in 1818. Mining experts believe that this element and its isotopes are abundant in the Earth's crust. Thorium is available as a metal, compound, or oxide. This element is primarily deployed in nuclear plant applications, because of lower cost and fewer cooling system components compared to uranium power nuclear reactors. Note that this metal comes in the form of a foil, rod, or sputtering target, while its compound comes in the form of sub-microns or nanopowder. It is important to mention that this material is widely deployed as a tungsten coating in electronic components due to its excellent emission characteristics. However, its fluoride and oxide are widely used in advanced electro-optical devices and components because of their high refractive index values. This rare earth material has been widely used in high-temperature glass applications such as in lamp mantels and in the development of crystal growth crucibles.

It is critically important to point out that thorium oxides are widely used for optical coatings and development of films using sputtering target. I must stress that its thin films are best suited for applications where performance, reliability, and longevity are the principal operational requirements. Note that its nanoparticles would yield ultra-high surface areas with unique properties best suited for scientific research and development activities.

Since thorium is a radioactive material, extreme care must be exercised in handling and transportation. Exposure to this material can lead to bone cancer and breathing this material could be lethal. It is important to mention that thorium isotopes are stable, which means non-radioactive, and are widely used in medical research studies. Studies performed on isotopes indicate that thorium has two isotopes, namely (Th-229) and (Th-230) with atomic mass of 229.0316 and 232.037, respectively. Note that these isotopes are available in abundance in the Earth's crust. Regarding thermal and electrical parameters, its room-temperature electrical conductivity is about 13 microohm-cm and its thermal conductivity is 54 W/cm-K. Due to its unique properties, nuclear scientist believe that the use of liquid fluoride in thorium-based nuclear power reactors will be found to be most useful and cost effective.

Comprehensive research and development studies undertaken by scientists during the 1950s and 1960s reveal that uranium-based power plants for generating electricity would be more expensive and it would require four to five years to complete the erection phase in comparison to thorium-based power plants. The studies further revealed that cooling the reactor, nuclear waste storage, and long-term radiation issues would be more expensive for uranium-based power plants. Because of these problems, nuclear scientists came to the following conclusions:

- Synthesis, structure, reactivity, and computational studies on thorium must be taken to comprehend the outstanding problems.
- Thorough scientific studies undertaken by nuclear scientists during the 1950s and 1960s show that uranium-based power reactors have more radiation, cooling, and storage of nuclear wastage problems compared to thorium-based nuclear power reactors. Furthermore, reactor cooling and radiation problems will be more severe. It is critically important to mention that uranium-based electrical power-generating plants will cost 4 to 5 million dollars and at least four years will be required to complete the erection phase depending on the electrical power-generating capability.

- Interaction of thorium with nitrate solution is critical and another solution is needed to avoid adverse solution problems.
- A cryogenic beam of refractory and chemically reactive molecules with expansion during cooling mode should be given serious consideration.
- Gamma-ray laser capable of emitting in the optical spectral range should be deployed in conducting optical experiments.
- Gamma-spectrometric analysis is simple and safe in thorium-based reactors.
- Background radiation levels must be kept as low as possible for the scientists and engineers.
- Medical-related problems due to reactor radiation level and dust concentration level analysis in non-coal mining must be undertaken to determine whether reactor radiation danger is more lethal than dust-related problems.
- Radiation exposure evaluation based on scientific measurements undertaken for occupational hygiene at research laboratories in Poland from 2001 to 2005 seems to indicate that exposure to fast neutrons in the vicinity of the nuclear reactor zone is most serious and scientific personnel must avoid such zones.

### 1.1.9 Properties and Applications of Gadolinium (Gd)

This radioactive REE was discovered in 1880. Its major application is in magnetic resonance imaging (MRI), which is widely used by doctors and physicians to locate problems with patients. It is important to point out that the high-resolution capability of the MRI equipment provides the doctors the exact location of a patient's bone fracture or acute medical problems in the heart or brain. Note that MRI equipment uses high magnetic fields which are associated with extreme noise levels based on magnetic core element vibrations. The noise level in the vicinity of MRI is extremely high and most of the patients do not like it. In summary, MRI findings are very reliable in medical assessment and treatment or diagnosis of a particular health problem.

Gadolinium is a critical component in computer tomography (CT) equipment, which plays an important part in computer-based medical technology. This technology is best suited for evaluation of congenital pulmonary vein abnormalities. Potential benefits of gadolinium and its isotopes can be summarized as follows:

- MRI detection of progressive cardiac dysfunction allows the attending physician to provide immediate and relevant care.
- It is interesting to mention that ultrasonography, computer tomography, and MRI techniques are best suited for specific medical treatment.
- Effectiveness of combined-enhanced MRI and contrast computer tomography techniques are best suited for most serious medical treatment.
- MRI is considered most appropriate in the assessment of distribution and evolution of mechanical dysynchrony in myocardial cases.
- Other techniques such as hyperpolarized spectroscopy are very effective in the detection of early stage tumors.

### 1.1.10  Characteristics of Isotopes of Gadolinium

Gadolinium was first discovered in 1880. This element is found in abundance in the Earth's crust. Its atomic structure, ionization energy, electrical conductivity, and thermal properties are best suited for MRI, medical diagnosis, and medical research instrumentation. It should be noted that this element is best suited as an injectable contrast agent for patients undergoing MRI procedure. Note MRI findings are very useful in the examination of the parotid gland. But MRI and computer tomography (CT) are also ideal for the evaluation of congenital pulmonary vein abnormalities. Note that the high magnetic moment capability of gadolinium can reduce the relaxation time, which can improve the signal intensity. This element can be used in X-ray cassettes, computer tomography, and other medical applications. Its other isotopes characteristics are summarized in Table 1.8.

**Table 1.8**   Isotopes of Gadolinium and Their Characteristics

| ISOTOPE | ATOMIC MASS | ABUNDANCE IN EARTH'S CRUST (%) |
| --- | --- | --- |
| Gd-152 | 151.920 | 0.21 |
| Gd-154 | 153.921 | 2.18 |
| Gd-155 | 154.922 | 14.80 |
| Gd-156 | 155.923 | 21.47 |

*1.1.10.1 Properties of Terbium (Tb)* This particular REE was first discovered by Carl Mosander in 1843. Comprehensive research studies indicate that this element is widely used in fluorescent lamps, phosphors, and high-intensity green emitters which are ideal for projection television sets. Terbium is available as a rare earth oxide, and as a compound with varying purity levels. This metal is available as wires, sputtering targets, foils, and pellets. Its compounds can be found as nano-particles and sub-particles. Its fluoride oxides are best suited for metallurgy:; chemical and physical deposition; and optical coatings. Its isotope Tb-159 has an atomic mass of 158.925 and is found in abundance in the Earth's crust. Its room-temperature electrical conductivity is 118 microohm and thermal conductivity is 11.1 W/meter-K.

Note that terbium is widely used as an X-ray phosphor. Terbium alloys are widely used in magneto-electric recording films. Latest research and development activities have analyzed the energy transfer between the terbium-binging peptide and red fluorescent proteins, the interaction between the metal ions and carbohydrates, the detection of bacterial endo-spores in soil, and the use of terbium complex as a luminescent probe for the imaging of endogenous hydrogen-peroxide generation in plant tissues.

*1.1.10.2 Thorium (Th)* This radioactive material was first discovered in 1818 and, with its isotopes, is found in abundance in the earth curst. This material is available as an element, as an oxide, or as a compound. Thorium is a lanthanide rare earth material, as shown in Figure 1.2, and is widely deployed in nuclear power plant applications. Note that a power plant can be designed for propulsion applications, widely used by destroyers, aircraft carriers, and steam turbines. It is important to mention that this material comes in various forms such as rod, foil, or sputtering target. It is widely deployed as a tungsten coating in electronic components due to its high-emission characteristics. Note that its fluoride oxide is best suited for advanced electro-optic components because of its high refractive indices. Thorium is widely used in high-temperature glass applications, e.g., in making mantels for portable high-intensity lamps. This REE is widely used to design crystal growth crucibles.

## 1.2 Summary

The history and properties of important REEs are described. Rare earth oxides and isotopes of critical rare earth materials are summarized with emphasis on their applications in medical, military, and commercial areas. Principal requirements for mining and processing of REEs and their oxides are discussed in great details. It is important to mention that methods for refining rare earth ores are described with particular emphasis on cost-reduction techniques, efficient techniques for screening, and refining methods. Important earth materials and oxides are briefly described. Critical properties of REEs and their oxides are summarized with emphasis on commercial, military, and medical applications. Various applications of rare earth alloys and compounds are identified with emphasis on cost and their suitability for specific applications. It should be stated explicitly that extraction of rare earth ores is strictly based on pre-knowledge of mineral deposits at specific locations. As stated before, the extraction of REEs, oxides, and alloys requires permission from the host country. It should be stressed that finding the appropriate location requires comprehensive metrological knowledge.

It is critically important to mention that rare earth materials found in mines and in the Earth's crust need to be refined and processed and must undergo quality-control inspections to satisfy the purchase requirements for specific applications. Estimating the expenses required for refining and processing REEs or oxides to meet specific performance specifications is the major issue for the purchasing agent. It is important to mention that processing REEs and oxides is very complex, time consuming, and expensive.

These costs can be reduced if the visual screening operator is has great expertise in selecting the precise hole size in the screen and the optimum screen inclination angle to accelerate the screening process without affecting the quality control process. This kind of processing will require minimum processing costs, if screen inclination angle and hole size are selected for optimum performance with no compromise in quality control and cost effectiveness.

The conversion of rare earth alloys and oxides into rare earth compounds is very complex and time consuming. For example, take a rare earth compound material such as samarium–cobalt, which is widely used in the development of samarium–cobalt magnets best suited for electric motors and generators. These rare earth magnets are widely employed in the design and development of hybrid electric and all-electric automobiles. They are preferred for such applications because they offer considerable reduction in size and weight and provide high thermal efficiency even when the operating temperatures exceed 200°C. It is important to mention that, in TWTAs deployed in fighter bombers, the operating temperatures can reach close to 300°C. Note that samarium–cobalt magnets offer a reduction in weight and size besides maintaining normal TWTA performance under high-temperature conditions, particularly during after-burner operations.

Research studies undertaken by the oxides reveal that certain rare earth oxides are best suited for specific commercial, military, and medical applications. Comprehensive research studies undertaken by the author on specific rare earth oxides indicate that thulium oxide and holmium oxide are best suited for infrared lasers. Radiation studies seem to indicate that greater radiation danger can be expected from the processed REEs than from the processed oxides. Mild radiation danger can be expected from rare earth isotopes, if they are not handled with extreme care. It should be mentioned that the radiation danger is strictly dependent on the half-life of the isotope. Doctors, nurses, and lab technicians must be familiar with the half-life of the rare earth isotopes, if they are using rare earth isotopes in medical treatment. This is absolutely essential to avoid irreversible health injury to the patient.

Specific REEs and oxides are identified for solid-state lasers capable of operating efficiently in infrared regions. Lithium material is widely used in the design and development of electric cells best suited for toys, cell phones, high-power portable sources for automobiles, electrical tools, heavy-duty wheelchairs, and trolleys to transport passengers in closed buildings and airport lounges. Rare earth oxides, namely zirconium oxides and yttrium oxide, are identified for fuel cell applications, because these oxides offer high fuel cell efficiencies and longer operating lives.

Comprehensive research studies on gadolinium oxide indicate that its oxides are considered ideal for optical coating and thin-film applications. Thin films of certain superconductive oxides such as YBSO (yttrium-barium-copper-oxide) and TBCCO (thallium-barium-calcium-copper-oxide) are best suited for microwave devices with minimum insertion loss at cryogenic temperatures. Some thin films of strontium titanate and magnesium oxide are most suitable for superconducting substrates, which can be used for deposition of thin films of MgO and $SrTiO_3$. It is important to mention that these superconducting substrates offer suitable critical temperatures and sintering temperatures needed for optimum microwave device performance at superconducting temperatures.

## Bibliography

A.R. Jha, *Rare Earth Materials: Properties and Applications*, CRC Press, Taylor & Francis Group, Boca Raton, FL, 2014 Edition, pp. 1–34.

A.R. Jha, *Fiber Optic Technology: Applications to Commercial, Industrial, Military and Space Optical Systems*, Scitech Publishing Company, Rayleigh, NC, 2007 Edition, p. 3.

A.R. Jha, *Superconductor Technology: Applications to Microwave Electro-Optics, Electrical Machines, and Propulsion Systems*, John Wiley & Sons, New York, NY, 1998 Edition, pp. 88–96.

A.R. Jha, *Next Generation of Fuel Cells and Batteries for Commercial, Military and Space Applications*, CRC Press, Taylor & Francis Group, Boca Raton, FL, 2012 Edition, pp. 86–90.

# 2

# DEPLOYMENT OF RARE EARTH MATERIAL IN THE REACTOR FOR ELECTRICAL POWER GENERATION

## 2.0 Introduction

Nuclear scientists predicted in the 1940s that a significant amount of electrical power could be generated from a nuclear fuel such as uranium-238 or plutonium-235. Nuclear scientists believe that more than 80% of the energy of fission appears as the kinetic energy of the fission fragments which are immediately converted into heat. The remaining 20% or so is liberated in the form of gamma rays and as kinetic energy of the fission neutrons. In summary, it can be stated that large amounts of energy can be generated with a very small amount of nuclear fuel. It is important to mention that the nuclear material must be fissionable material. It is critically important to point out that readers should be able to convert the fission energy values into practical units. It should be noted that 1 million electron vols (MeV) is equal to $1.60 \times 10^{-6}$ ergs, which is equivalent to $1.60 \times 10^{-6} \times 10^{-7}$ watts-second or joules. This means 1 MeV is equal to $1.60 \times 10^{-13}$ watts-second or equal to joules. It should be mentioned that the total energy produced by the nuclear reactor per fission is roughly equal to 200 MeV, which is equal to about $3.2 \times 10^{-11}$ watts-second or joules. In other words, to generate one watt of electrical power, fissions at the rate of $3.2 \times 10^{10}$ per second are required, provided that the nuclear reactor has been operating for some time to meet the fission rate requirement.

Readers are advised to get familiar with terms such as fission and individual nuclei. It should be noted that the atomic weight of a nuclear element is expressed in grams. Furthermore, a nuclear element contains the Avogadro's number, which is equal to $6.02 \times 10^{23}$, of individual nuclei, and if all these nuclei undergo fission, the electrical energy liberated will be close to $(6.02 \times 10^{23})$ $(3.2 \times 10^{-11})$, which boils down to $1.9 \times 10^{13}$ watts-second or joules or $5.28 \times 10^6$ [1, 2] kW hour or 5.28 million electrical units, because one electrical unit is defined as 1 kW of electrical energy consumed in one hour. In other words, it can be stated that $5.28 \times 10^6$ kW hour or 5.28 megawatt hour of heat generated by the complete fission of 233 grams of uranium-233 or 235 grams of uranium-235 or 239 grams of plutonium-239. It is important to mention that besides the generation of electrical energy, the fissionable materials, as mentioned, undergo other nuclear reactions such as the heat liberated per unit weight of the fissionable rare earth material. For all practical purposes, it can be stated that the power production per day by 1 gram of fissionable material would be roughly $10^6$ megawatt and instant release of heat due to fission process can be summarized in Table 2.1.

Important elementary features can be summarized as follows:

Total instantaneous heat generated = 185 MeV, with 90% of heat liberated instantly.
Total delayed fission products from beta particles (7) and gamma rays (6) amount to 13 MeV.

It should be stated that radiation from products captured could be around 2 MeV. Furthermore, it can be stated that, in the course of time, as the fission products and capture products accumulate and decay, the overall heat level will increase to 200 MeV roughly. As mentioned above, the total heat available per fission will be 200 MeV. Note that this level will remain constant, provided that the nuclear reactor is in equilibrium condition [3]. The author has summarized the important conversion units in Table 2.2 for the benefit of the readers.

It is important to point out that actual photon energies must be considered to comprehend certain aspects of reactor shielding and

**Table 2.1**    Instant and Appropriate Release of Heat due to Fission Process (MeV)

| | |
|---|---|
| Energy of fission fragments | 168 |
| Energy of fission neutrons | 5 |
| Instantaneous gamma rays | 5 |
| Capture gamma rays | 7 |

**Table 2.2**   Conversion of Units Used in This Book

| TO CONVERT | INTO | MULTIPLY BY |
|---|---|---|
| 1 MeV | erg | $1.60 \times 10^{-6}$ |
| 1 MeV | watts-second | $10^{-7}$ |
| 1 gram per fission | kW hour | $5.3 \times 10^{6}$ |
| 1 watt | erg | $10^{7/sec}$ |
| 1 watt | BTU/min | 0.0569 |
| 1 watt | Ft-lb/min | 44.26 |
| 1 kW hour | 1 unit | 1 |

heat removal. Note that experimental studies should be conducted on the nuclear energy distribution to capture gamma radiations. The capture of gamma-ray energies of reactor materials is summarized briefly in Table 2.3.

It should be mentioned that among the substances listed in the above-mentioned table, only with hydrogen and beryllium is it possible to capture gamma radiation emitted as a single photon per neutron capture. Note that most of the energy is associated with fast neutrons, whereas relatively low energy is associated with slow neutrons. It should be stressed that slow-neutron reactions accompanied by the emission of a charged alpha particle or a proton are rare. It is critically important to mention that when uranium-235 is involved with a reactor core, it can take part in fission as well as in radiative capture.

**Table 2.3**   Capture of Gamma-Ray Energies of Critical Reactor Materials (MeV)

| REACTOR MATERIAL | PHOTON ENERGY (MeV) | ENERGY PER CAPTURE (MeV) |
|---|---|---|
| Aluminum | 7.7 | 2.7 |
| | 6.0 | 1.4 |
| | 3.0 | 3.8 |
| Beryllium | 6.8 | 6.8 |
| Carbon | 1—3 | 0.6 |
| | 5.0 | 5.5 |
| Hydrogen | 2.2 | 2.2 |
| Lead | 6.7 | 0.5 |
| | 7.4 | 6.8 |
| Uranium | 3.0 | 6.0 |
| | 1.0 | 0.8 |
| Zirconium | 3–5 | 4.5 |
| | 5—7 | 2.0 |

The term "neutron absorption" is used to include both reactions, whereas "neutron capture" refers to the latter only. In summary, it can be stated that in a thermal reactor, where fast fission neutrons are being produced while the slow neutrons are readily absorbed, there are positive deviations from the Maxwell–Boltzmann distribution at high energy and negative deviations at low energy [1].

## 2.1 Kinetic Energy of Thermal Neutrons

It should be noted that thermal neutrons have very small energy and others have very large energies. Note that the kinetic energy per unit velocity interval is strictly based on the Maxwell–Boltzmann distribution. The Boltzmann constant (k) can be written as:

$$k = \left[ 1.38 \times 10^{-16} \right] \text{ergs per } °C \qquad (2.1)$$

OR

$$= \left[ 8.6 \times 10^{-5} \right] \text{electron} - \text{volt per } °C \qquad (2.2)$$

OR

$$= \left[ 4.8 \times 10^{-5} \right] \text{electron} - \text{volt per } °F \qquad (2.3)$$

The expression for the kinetic energy can be written as:

$$KE = \left[ 1/2\,m\,v^2 \right] \text{electron volt} \qquad (2.4)$$

where, m is the mass $= [1.67 \times 10^{-24}]$ gram and velocity parameter (v) is defined as cm/sec. Therefore the velocity parameter can be finally written as:

$$v = \left[ \left(1.4 \times 10^6\right)\left(E^{0.5}\right) \right] \text{cm} / \text{sec} \qquad (2.5)$$

where E is the neutron energy expressed in electron volts.

Using the result obtained by the above Equation (2.5) and combining with the expressions for the thermal-neutron energy as derived above, the average speed of the thermal neutrons can be given as,

$$= \left[ \left(1.3 \times 10^4\right)\left(T_K\right)^{0.5} \right], \qquad (2.6)$$

where $T_K$ is the absolute temperature in Kelvin
   OR

$$T_R = \left[ (0.97)(10^4)(T_R)^{0.5} \right],  \qquad (2.7)$$

where $T_R$ is the absolute temperature in Rankine cycle.

### 2.2 Neutron Reactions and Their Production in the Nuclear Reactor

Research studies performed on the subject concerned indicate that neutrons undergo two types of reactions with atomic nuclei that are characterized as absorption (or capture), in which a neutron enters the nucleus, and scattering, in which the neutron interacts with, and transfers some or all its energy to, the nucleus, but the neutron remains free after the process. As stated in the previous chapter, both kinds of interaction play a critical role in the operation and control of nuclear reactors.

In considering absorption reactions it is convenient to distinguish between the reactions of slow and fast neutrons. It should be noted that there are four main kinds of slow and fast neutrons; these involve capture of the neutron by the target followed by the emission of gamma radiation; or by the ejection of an alpha particle; or by the ejection of a proton; or by fission. Note that among these, the radiative capture process is the most common because it occurs with a wide variety of elements. It should be noted that number 2 and 3 reactions with slow neutrons are limited to a few isotopes of low mass numbers, whereas fission by slow neutron is restricted to certain nuclei of high atomic number [3].

It is important to point out that in a thermal reactor, where fast fission neutrons are being produced, the slow neutrons have been already absorbed. This creates positive deviations from the Maxwell-Boltzmann distribution at high energies and negative deviations at low energies.

### 2.3 Critical Properties of Rare Earth Elements and Emitters Produced during the Fission

Critical properties of rare earth elements and emitters produced during the fission in the rector can be briefly summarized as follows with specific comments:

- It is critically important to point out that the action of gamma rays of moderate energy (say 0.5 MeV) level on deuterium (or heavy water) and beryllium will yield mono-energy neutrons.
- Slow neutrons are available from an alpha-emitting source such as a mixer of radium or polonium of an alpha emitter.
- It is critically important to remember that neutrons produced by the interaction of alpha particles and beryllium have fairly high energy range, ranging from 5 to 12 MeV or more.
- Rough calculations performed by the author seem to indicate that 1 gram of pure uranium-238 or thorium-232 during the fission would produce roughly $10^6$ watts or 1 megawatt of electrical energy.
- These preliminary calculations further indicate that these fissionable materials would require at least MeV of energy to cause fission. In summary, it should be stated that there must be enough neutrons of sufficient energy to maintain the chain reaction in the nuclear reactor to produce electrical energy.
- Calculations performed for heat generation reveal that complete fission of 233 grams of uranium-233, or 235 grams of uranium-235, or 239 of grams of plutonium would be needed to generate heat energy equal to $(5.3 \times 10^6)$ kW hour or (5.3) megawatt hour.
- Note that $(2.8 \times 10^{13})$ ft-lb of heat energy can be generated by 1 pound of fissionable material. It is essential to mention the conversion factors between the electrical and thermal parameters. Note for example that 1 watt equals 0.0569 BTU/min and 1 erg/sec is equal to 10 megawatts.

### 2.4 Description of the Critical Elements of a Nuclear Reactor

Critical components of a nuclear reactor can be seen in the reactor core, steam turbine, heat exchanger (or oiler), cooling water line from condenser, and control rods located in the core are clearly shown in Figure 2.1. Note that an enormous amount of heat is generated in the core during the fission process. Due to this, all reactors are known as thermal reactors. The core consists of fissionable material in the fuel rods and moderator for controlling the energy of neutrons. The nuclear core is surrounded by a reflector of suitable material. It is important to

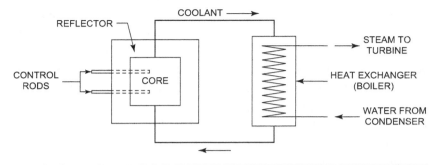

**Figure 2.1**    Critical components of nuclear power reactor.

mention that the core and the moderator are capable of maintaining the fission chain reaction at a specified power level. Control of fission reaction is achieved either through control rods containing a strong neutron absorbing material or through neutron shielding or reflectors in order to prevent the escape from the reactor core. A thermal shield and a radiation shield surround the reactor for the protection of the operating personnel or research scientists working in the vicinity of the nuclear reactor installation. Absolute control of the nuclear reactor is necessary throughout its operation. The safety of the reactor is of critical importance. If the reactor is operating at full power rating, then excessive heat is generated. For the safety of the reactor and its associated accessories, the heat produced by the fission must be removed as quickly as possible. This requires the most efficient coolant to circulate throughout the reactor in order to maintain a safe temperature distribution that is as uniform as possible. According to nuclear power designers and scientists, the main focus must be on the cooling of the reactor core to avoid overheating of the core. In other words, the problem of heat transfer from the core squarely depends on the rate of heat removal and the temperature of the cooling water or the coolant. It should be remembered that the rate of heat removal from the core determines the power output of the nuclear power plant. The method used for cooling the reactor core is of fundamental importance. Comprehensive research studies by the author seem to indicate that inadequate cooling and overheating of nuclear reactors has been observed over 30 years in more than five technically advanced countries around the world. The overheating of nuclear reactors is a classical problem and it must be avoided under all operating conditions.

Nuclear scientists and engineers are considering a most radical and unproven operating concept. They are considering using the reactor directly as a boiler to vaporize the water, leading to conversion into a superheated steam to drive the turbine, which is mechanically coupled to an electrical generator-exciter set. This would lead to production of electricity. The design of boiling reactors was under investigation during the 1960s and 1970s. There is no record of successful operations of such boiling reactors (Figure 2.1).

It should be noted that in most power reactor designs, the heat energy is taken up by a circulating coolant or water, which transfers it to a heat exchanger. The superheated steam generated is used to drive a high-capacity alternator or electric generator. Note that a heat exchanger associated with a nuclear reactor plays a role as a boiler in a normal power plant. In summary, significant improved reliability, ultra-high thermodynamic efficiency and rapid heat removal from the reactor core are the principal design requirements for nuclear power reactors.

### 2.4.1 Operational Requirements for Nuclear Reactor Materials

Comprehensive research studies undertaken by the author reveal that thorium, uranium-238, and uranium-235 are the most desirable reactor core materials to produce a large output of power. Since enormous heat is generated at high temperatures within the reactor core, reactor structural materials, thermal reflector materials, shielding materials, and the type of coolant must be selected to provide optimum safety and reactor efficiency throughout the fission duration. Not only are these materials able to withstand high operating temperatures exceeding 1000°C or more, but they also have low cross sections for neutron absorption. Note that neutron absorption is the most essential requirement for the reactor core and accessories. Nuclear reactor design engineers and scientists insist that any loss of neutrons must be compensated by an increase in the amount of fissionable material such as thorium and uranium if a chain reaction is to be maintained at all times in the reactor zone. Maintenance of chain reaction is absolutely necessary in any nuclear reactor regardless of application.

It is important to mention that neutron bombardment in a reactor causes permanent displacement of the atoms from their normal positions in the crystal lattice structure. This could lead to a disruption in the internal structure, which could change the physical properties of the fission materials of the reactor. For the safety of the reactor, any change in the physical properties requires appropriate changes for the structural materials as well as in the design of the fuel elements.

The coolant used in the reactor must not absorb neutrons and it must be stable at high temperatures and in the presence of high neutrons and gamma radiation fields. In the presence of these high radiation fields, even organic cooling liquids cannot meet such stringent performance specifications.

### 2.4.2 Materials for Nuclear Fuel, Moderator, Reflector, and Thermal Shield

Critical components for a nuclear reactor include nuclear fuel, moderator, reflector, and thermal shield. As mentioned earlier, potential nuclear fuel can employ only selected rare earth materials such as thorium, uranium-232 or plutonium-235. The nuclear fuel is inserted in the rods of appropriate fissionable material. The rods can be inserted into the core to control the amount of fissionable material and maintain the operation of the nuclear reactor. A large amount of fuel material is required initially for the reactor in case of natural uranium-238, which can range from 25 tons to 55 tons. However, in case of highly fissionable fuels such as plutinium-235, significantly smaller amounts of fuel material may be needed, ranging from 7 pounds to 12 pounds when the reactor is started.

**Moderator materials** include regular water, heavy water, and graphite. The amount of moderator material depends on the fuel requirement and other factors that have control over the reactor efficiency and the amount of moderator needed to sustain the atomic change reaction.

**Reflector materials** are needed to reduce the reflections from the reactor core. Several materials and technologies are available which to cut down the radioactive reflections. Materials such as graphite, a rare earth material (2 ft thickness), can circulate regular water and heavy water in the quantity of 5,000 to 30,000 cu/ft heavy flow per minute.

**Cooling water** is circulatory regular water and circulatory heavy water that are best suited for coolant mediums with flow rates ranging from 5,000 cu/ft per minute to 30,000 cu/ft per minute. Closed-cycle cooling technology must be deployed to retain high cooling efficiency with minimum loss of water.

**Shielding materials** are deployed to prevent proton loss from the reactor. Shielding materials include concrete, special concrete, lead, cast iron, steel, and a combination of concrete (5 ft thick on all sides plus 4 ft of wood on all sides) and lead layer on the top. Sometimes, a combination of lead and cast iron is used for moderate shielding capability where a reduction in shielding cost is of paramount importance, particularly for low-power nuclear reactors [3].

### 2.5 Fission Products Produced in the Reactor and Their Properties

Fission products are generated in the nuclear reactors after they have been operating for some time. Experiments conducted by nuclear scientists reveal that isotopes make the greatest contributions to the radioactivity of fission products. These isotopes have potential applications in the medical field and in commercial discipline. The best known fission products include cesium (Ce-137), strontium (Sr-90), praseodymium (Pr-144), yttrium (Y- 91), cerium (Ce-144), strontium (Sr-89), cerium (Ce-141), lanthanum (La-140), praseodymium (Pr-143), barium (Ba-140), iodine (I-133), and others of lesser significance. Fission products with long half-lives must be carefully handled, because they might have residual radioactivity. Note that fission products emitting beta particles of low energy or less than 1 MeV do not present a serious hazard, but they should still be handled carefully during storage and transportation. It should be noted that no gamma radiation can be ignored. One role of thumb states that when the half-life is less than 24 hours, the fission products or their isotopes can be handled with minimum danger. Studies performed on fission products indicate that during the reactor cooling period, the isotope iodine-131 with an eight-day half-life cooling period can still be dangerous in terms of radiations if present in the gases released from the processing plant, because it undergoes considerable decay. There can be radiation danger from some isotopes even after the expiration of their half-life periods. The author summarizes the half-lives of isotopes that make major contributions to the radioactivity of the fission products in Table 2.4.

The half-lives in the above table are expressed in "Curie". Note 1 curie is equal to $3.7 \times 10^{10}$ interactions per second. After 100 days of cooling, the radiation levels in the isotopes can be as low as a few microcuries, which are not considered dangerous for laboratory research personnel.

**Table 2.4**  Typical Half-Lives of Radioactive Fission Products

| FRACTION OF TOTAL CURIE LEVEL OF ACTIVITY | CURIE LEVEL AFTER COOLING TIME (DAYS) | | | | |
|---|---|---|---|---|---|
| | 20 | 40 | 60 | 80 | 100 |
| Cerium-137 | 0.13 | 0.14 | 0.21 | 0.25 | 0.28 |
| Strontium-90 | 0.11 | 0.12 | 0.16 | 0.20 | 0.22 |
| Praseodymium-144 | 0.25 | 0.30 | 0.28 | 0.22 | 0.31 |
| Yttrium (Y-91) | 0.07 | 0.05 | 0.08 | 0.04 | 0.01 |
| Lanthanum (La-140) | 0.03 | 0.01 | 0.005 | 0.002 | 0.001 |
| Barium (Ba-140) | 0.02 | 0.01 | 0.005 | 0.003 | 0.001 |
| Iodine (I-131) | 0.007 | 0.001 | 0.0003 | 0.00001 | 0.000005 |

It is critically important to point out that the effect of the infinite multiplication factor is strictly dependent on the interactions of both the reactor fuel and the moderator, and on their relative proportions. It should be further mentioned that the ratio of fissionable material-to-moderator will determine the neutron energy range within which the majority of the fissions occur. In case there is very little moderator in the reactor, the coolant can act as a moderator and, therefore, most of the fissions will be caused by the absorption of fast neutrons by the fissionable nuclei. Such a nuclear reactor is classified as "thermal reactor". When a large amount of moderator is used in a reactor, slow neutrons or thermal neutrons will be responsible for most of the fission reactions. Nuclear scientists claim that most of the rectors now in operation are thermal reactors.

## 2.6 Critical Operational Status of Nuclear Reactor

When the status of the reactor becomes "CRITICAL", then output power become available Assuming "k" is the number of neutrons produced in each generation, and only one of these neutrons is required to maintain the fission reaction chain, when the reactor is "critical" and power output is available.

## 2.7 Nuclear Reactor Operation Using Various Fuels, Moderators, and Coolants

Comprehensive research studies performed by the author reveal that the choice of moderator and coolant depends on the use of nuclear fuel

**Table 2.5**   Summary of the Tradeoff Studies Conducted on Reactors

| REACTOR FUEL USED | MODERATOR | COOLANT | NEUTRONS TYPE | DIMENSIONS(FT) |
|---|---|---|---|---|
| Natural uranium | Graphite | Air | Slow | 25 |
| Natural uranium | Heavy water | Heavy water | Slow | 10 |
| Slightly enriched U | Graphite | Sodium | Slow | 8 |
| Slightly enriched U | Water | Water | Slow | 5 |
| Highly enriched U | Water | Water | Slow | 1.5 |
| Highly enriched U | None | Sodium | Fast | 1 |

and on a restriction on the size of the reactor. Based on these studies one can conclude that a critical reactor deploying natural uranium as a reactor fuel with heavy water as a moderator will achieve significant reduction in the overall size of a reactor. Results of tradeoff studies conducted by nuclear scientists are summarized in Table 2.5 [2].

## 2.8 Radioactivity of Fission Fragments in the Reactor

According to nuclear scientists, practically all the fission fragments produced in a nuclear reactor are radioactive, as they emit negative beta particles. Scientists further believe that the immediate decay products are mildly radioactive, because each fission fragment is followed by three stages of decay before a stable species is formed. The studies performed by the author indicate that there are several radioactive species present among the fission products after a short time.

Note that large number of fission products emit gamma rays besides beta particles. They are known as delayed mission gamma radiation. It is important to mention that most of these contain moderate energy not exceeding 2 MeV. High-energy gamma rays generated from the mission products including their isotopes are summarized in Table 2.6 [3]. Note only high percentages of gamma yield are given priority.

Fission products with low fission yields and low gamma yields shown in the above-mentioned table are not included in Table 2.5. Furthermore, it should be noted that the gamma yield is proportional to the decays that are accompanied by gamma radiation in each case. It should be further noted that the product stated in percentage in the third and fourth columns offers the number of gamma photons of each particular kind per 10,000 fissions.

**Table 2.6** Summary of High-Energy Gamma Rays from Fission
Products along with Isotopes

| RADIOACTIVE SPECIES | HALF-LIFE (SEE NOTE) | FISSION YIELD (%) | GAMMA YIELD (%) | ENERGY (MEV) |
|---|---|---|---|---|
| $Ce^{144}$ | 290 days | 5.3 | < 1 | > 2.2 |
| $Pr^{144}$ | 18 min | 5.3 | < 1 | > 2.2 |
| $La^{140}$ | 40 hour | 6.1 | 3.2 | 2.5 |
| $Te^{132}$ | 77 hour | 4.5 | 2.7 | 2.0 |
| $I^{132}$ | 2.4 hour | 4.5 | 2.7 | 2.0 |
| $I^{135}$ | 6.7 hour | 5.6 | 1.9 | 2.4 |
| $Kr^{88}$ | 2.8 min | 3.1 | 15 | 2.8 |
| $Rb^{88}$ | 18 min | 3.1 | 26 | 1.9 |

## 2.9 Estimate of Rate of Beta and Gamma Energy Release by Fission Products after the Reactor Shutdown

Complete shutdown of the nuclear reactor is absolutely necessary before any maintenance or repair can be undertaken. After the reactor shutdown, a cooling process must begin, and it could take several weeks before any maintenance or repair can be undertaken. Following these activities, the level of radioactivity must be measured in the core, cooling water, condenser tubes, and any other items directly involved in the operation of the reactor. Maintenance or repair activities can start with great caution, if the radiation levels from all sources including beta radiation, gamma radiation, and other sources are within the safe and permissible limits.

Theoretically, it is possible to determine the rate of decay of the complex fission product mixtures in terms of fission yields and the radioactive constants or parameters of the various isotopes of the reactor and the radioactive constants of the various isotopes. Exact determination of the fission decay with reasonable accuracy will be extremely difficult if not impossible. Finally, the plant operator has to face two serious difficulties: first to achieve the estimate rate of heat energy release from the fission products due to emission of beta particles and gamma rays, and second to determine the amount of a particular fission product present in the reactor at various times. Note that the half-lives of the fission products from a rector typically range from a few seconds to millions of years. Nuclear scientists state that the rate of emission of beta particles and gamma rays can be determined by means of empirical expressions [4] that could be accurate within a factor of two or less.

The following empirical expressions yield reasonably fair estimates after 10 seconds up to several weeks after the fission has taken place. These empirical expressions can be written as follows:

Rate of emission from beta radiation

$$= \left[ \left( 3.8 \times 10^{-6} \right) \left( t^{-1.2} \right) \right] \text{particles} / (\text{sec.})(\text{fission}) \tag{2.8}$$

Rate of emission from gamma radiation

$$= \left[ \left( 1.9 \times 10^{-6} \right) \left( t^{-1.2} \right) \right] \text{photons} / (\text{sec})(\text{fission}) \tag{2.9}$$

where t indicates the time after fission in DAYS. Note that day is equal to $(24 \times 60 \times 60) = [86,400]$ sec.

Studies performed by the author on the subject reveal that the mean energy of beta particles from the fission products is roughly 0.4 MeV, while that of gamma-ray photons is about 0.7 MeV. Using the mean energy conditions, Equation (2.8) can be rewritten as

Rate of emission of beta and gamma energy

$$= \left[ \left( 2.66 \times 10^{-6} \right) \left( t^{-1.2} \right) \right] \text{Mev} / (\text{sec})(\text{fission}) \tag{2.10}$$

It should be noted that t is the time in days after fission activity started in the reactor.

Cooling time calculations for a reactor operating after the fission over T days [5].

Assuming a fission activity occurring during whole time "Tau" when the reactor is operating.

Assuming $T_0$ = time over which a reactor operates with constant power output at level P watts, consider a small interval "dT" at a time T after the start-up as illustrated in Figure 2.2.

$T_0$ = time after reactor operating
Tau = the time after start-up (tau) due to fission occurring
[Tau − $T_0$] = time in days after the reactor shutdown or reactor cooling time

One must consider the rate of emission of beta and gamma energy at a time "tau" after the start-up due to fissions occurring during the time interval dT, which can be written as follows:

$$\left[2.70 \times 10^{-6}\right] \left[\text{Tau} - \text{T}\right]^{-1.2} \text{Mev} / (\text{sec})(\text{fission}) \qquad (2.11)$$

$$\text{Note}\left[3.1 \times 10^{10}\right] \text{fissions} / \text{sec yield 1 Watt of power} \qquad (2.12)$$

$$\text{Note}\left[3.1 \times 10^{10}\right]\left[24 \times 60 \times 60\right] = \left[3.1 \times 10^{10} \times 86,400\right]$$
$$= \left[2.678 \times 10^{15}\right] \text{fissions} / \text{day yield 1 Watt} \qquad (2.13)$$

During the interval dT, the reactor is operating P watt/day with a number of fissions occurring per day. The rate of emissions of beta and

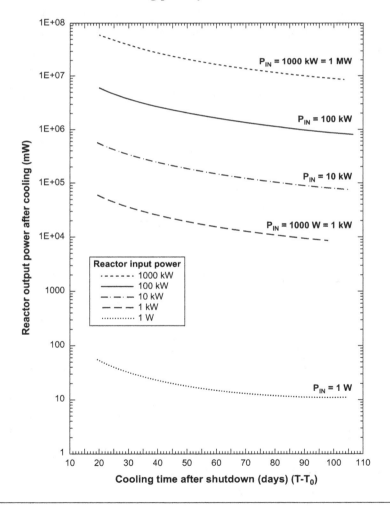

**Figure 2.2** Cooling time calculation for reactor power output over cooling time after reactor shutdown.

gamma energy at a time "tau" due to fissions in the time interval "dT" can be written as,

$$\frac{\left[2.7\times10^{-6}\right]\left[2.68\times10^{15}\right]\left[P\times dT\right]}{\left[Tau-T\right]^{-1.2}} \quad Mev/sec \tag{2.14}$$

OR

$$\left[7.24\times10^{9}\right]\left[P\right]\left[dT\right]\left[Tau-T\right]^{-1.2} \quad Mev/sec \tag{2.15}$$

Considering the fissions occurring for the entire period of reactor operation from time $T=T_0$ to $T=0$, the rate of emissions of beta and gamma energy at time "tau" can be expressed as

$$=\left[\left(7.24\times10^{9}\times P\right)/^{0}_{T_0}\left(Tau-T_0\right)^{-1.2}dT\right] \tag{2.16}$$

$$=\left[3.7\times10^{10}P\right]\left[\left(Tau-T_0\right)^{-.2}-Tau^{-.2}\right] \quad Mev/sec \tag{2.17}$$

As mentioned earlier that 1 Mev

$$=\left[1.60\times10^{-13}\right] \text{ watt sec}\left(\text{note}:1\,Watt=10^{-7}erg.sec\right) \tag{2.18}$$

This means 1 Mev $=\left[1.60\times10^{-6}\times10^{7}\right]$ watt $-$ sec

$$=16\,watt-sec=joules \tag{2.19}$$

Inserting Equation (2.19) into Equation (2.17), we get,
    Rate of emission of beta and gamma energy at time "tau", we get

$$\left(5.92\times10^{-3}\right)\left(P\right)\left[\left(Tau-T_0\right)^{-.2}-Tau^{-.2}\right] \tag{2.20}$$

It should be noted that P is the reactor output power in watts and [Tau $-T_0$] is the time in days after **reactor shutdown, which is also known as the cooling time for the reactor.** Equation (2.20) offers computed reactor power output as a function of cooling time and the results are shown in Table 2.7.

**Table 2.7** Summary of Reactor Power Output as a Function of Cooling Time

| $(TAU - T_0)$ | $(TAU - T)^{-2}$ | $(TAU)^{-2}$ | $[(TAU -T)^{-2} - (TAU - T)^{-2}]$ | POWER OUTPUT (WATT) |
|---|---|---|---|---|
| 20 | 0.01 | 0.00055 | .00945 | $10^{-6} \times 55.75$ |
| 40 | 0.005 | .00055 | 0.00445 | $10^{-6} \times 26.25$ |
| 60 (2 mo) | 0.00333 | 0.00055 | 0.00278 | $10^{-6} \times 16.40$ |
| 80 | 0.00250 | 0.00055 | 0.00195 | $10^{-6} \times 11.50$ |
| 90 (3 mo) | 0.00222 | 0.00055 | 0.00167 | $10^{-6} \times 9.85$ |
| 100 | 0.00200 | 0.00055 | 0.00145 | $10^{-6} \times 8.55$ |
| 120 (4 mo) | 0.00167 | 0.00055 | 0.00112 | $10^{-6} \times 6.58$ |

*REMARKS:* These calculations reveal that the reactor power output will be about 56 microwatts after 20 days' cooling, 16.4 microwatts after two months' cooling, about 9.8 microwatts after three months' cooling, and down to about 6.6 microwatts after four months' cooling. These calculations further reveal that the cooling of the reactor is progressively improving and the reactor output drops to the level of a few microwatts, which is not dangerous and relatively radiation-free. In other words, the total beta-particle and gamma-ray power levels from the fission products are safe and could be removed after four months of cooling using ordinary water as a coolant. Note that ordinary water from a river or the ocean can meet the cooling requirement with minimum cost provided it is free from suspended impurities and, especially, from dangerous chemical solutions. This is to keep cooling tubes free from impurity deposits. A plot of the calculated results can be seen in Figure 2.2.

## 2.10 Most Serious Maintenance Problems Observed in Reactors around the World

The author has reviewed some published maintenance reports on nuclear power reactors and has summarized the various problems due to poor or inadequate maintenance and inappropriate repairs. The maintenance data on reactors operating around the world have been summarized in Table 2.8, pointing out the most serious problems that can force a reactor shutdown over several months because of extensive repair work requirement. Comprehensive research studies undertaken

**Table 2.8** Most Serious Cases of Nuclear Power Rector Accidents and Shutdowns

| YEAR OF ACCIDENT | COUNTRY INVOLVED | MAJOR CAUSES OF REPAIR OR SHUTDOWN |
|---|---|---|
| 2011 | Fukushima (Japan) | Total shutdown due to fire and earthquake |
| 2011 | Onagawa (Japan) | Reactor shutdown due to excessive fire |
| 2006 | Belgium | Severe health risks due to high levels of RA |
| 2006 | Sweden | Reactor shutdown due to emergency power failure |
| 2006 | US | Highly enriched uranium solution leakage |
| 2002 | Canada | Excessive heat due to coolant loss |
| 1999 | France | Leakage of coolant, wastewater contamination |
| 1986 | Chernobyl (Russia) | Excessive heat, melt-down of core and shutdown |

**Table 2.9**   Information on Worldwide Reactor Accidents, Shutdowns, and Radioactive Leaks

| YEAR | LOCATION (NO OF REACTORS) | CAUSES OF ACCIDENTS, SHUTDOWNS, AND RADIOACTIVE LEAKS |
|---|---|---|
| 1986 | Germany (3) | Failure of circulatory water pump, fuel elements damaged |
| 1987 | | |
| 1999 | | |
| 1987 | India (6) | Reactor shutdown due to refuel accident, leakage of radioactive iodine and helium, and heavy water reactor melt-down. Repairs took two years. |
| 1975 | Japan (19) (see below) | The reactor was out of commission for two years. Radioactive leaks happened five times. Other incidents include a fire due to a steam turbine blade accident, a radioactive wastewater leak, a sodium radiation leak, a fire accident due to failure of fuel-loading system, an explosion due to poor maintenance, a melt-down due to overheating of the core, a reactor shutdown due to heavy leakage of coolant, and several instances of radiation leakage. |
| 1978, 1979, 1980, 1985, 1986, 1991, 1995, 1997, 1999(June), 1999 (Sept), 2002, 2004, 2007 2009 (March), 2009(Dec), 2011 (March), 2011 (October) | | |
| 2011 (October) | Pakistan (1) | Kanupp nuclear plant imposed a seven-hour emergency due to heavy water leakage from a feed pipe line to the reactor. |
| 1992 | Russia (3) | A MBMR reactor released a radioactive cloud that travelled in a north-east direction. |
| 1997 | Leningrad | Due to cracks in the wall, a ground water leak was observed at the wastewater storage facility. |
| 1998 | Leningrad | A RBMK reactor was shut down due to a radioactive leak. |
| 1999 (October) | South Korea | A steam explosion was responsible for a shutdown; a leakage of radioactive liquid gas meant that 12 gallons of heavy water leaked during the pipe maintenance and 12 people were exposed to severe radiation during the clean-up. |
| 1986 (April) | Ukraine (2) | Steam explosion caused melt-down and the forced evacuation of 300,000 people due to a heavy radioactive cloud. |
| 1999 (October) | | A metal structure collapsed causing a gamma-ray source to fall out of its container, which caused high levels of radiation leak, and the reactor was shut down for safety. |
| 1957 (March) | England (6) | Intense radioactivity was released which contaminated farms with high levels of strontium-90. |
| 1957 (October) | England | A fire ignites plutonium piles causing high levels of contamination. |

*(Continued)*

**Table 2.9 (Continued)**   Information on Worldwide Reactor Accidents, Shutdowns, and
Radioactive Leaks

| YEAR | LOCATION (NO OF REACTORS) | CAUSES OF ACCIDENTS, SHUTDOWNS, AND RADIOACTIVE LEAKS |
|---|---|---|
| 1967 (May) | Scotland | Partial melt-down of the reactor. Graphite moderator particles blocking fuel channel element to the reactor. |
| 1996 (Sept.) | Scotland | Fuel processing plant shutdown due to intense radiation level in the wastewater going to the sea. |
| 1998 (Feb.) | England | High radiation level due to a leak from a damaged bag containing a nuclear filter. |
| 2005 (April) | England | 20 tons of uranium and 160 kg of plutonium leaked from a cracked pipe at the fuel processing plant resulting in massive radioactive radiation levels. |
| 1955 (Nov.) | Idaho Falls (US) (38) | Partial reactor core melt-down. |
| 1959 (July) | Simi valley (US) | Core melt-down in a reactor using sodium as coolant. |
| 1961 (Jan.) | Idaho Falls (US) | Explosion due to overheating in the reactor. |
| 1966 (October) | Michigan (US) | Sodium cooling system malfunction causing melt-down. |
| 1973 (August) | Michigan (US) | Steam leakage forcing shutdown of pressurized water reactor. |
| 1975 (March) | Alabama (US) | Fire caused damage to control cables, which interrupted the water flow to the reactor. |
| 1975 (Nov.) | Nebraska (US) | Hydrogen gas explosion caused fires, which shutdown the boiling water reactor. |
| 1977 (June) | Connecticut (US) | Hydrogen gas explosion caused shutdown of pressurized water reactor. |
| 1979 (Feb.) | Virginia (US) | Reactor shutdown due to failing tube bundles in steam generator. |
| 1979 (March) | Pennsylvania (US) | Loss of cooling caused partial core melt-down. |
| 1981 (Oct.) | New York (US) | 100,000 gallons of Hudson River water leaked into Indian Point Energy Center and flooding was noticed up to 9 ft in the reactor. |
| 1982 (March) | New York (US) | Recirculation pipes failed causing reactor shutdown for two years for clean-up. |
| 1982 (April) | New York (US) | Damage to steam generator tubes caused main generator shutdown for more than one year. |
| 1982 (June) | S. Carolina (US) | Feed water heat extraction line fails causing pressured water reactor damages to thermal cooling system. |
| 1983 (Feb.) | New Jersey (US) | Safety inspection fails causing two-year shutdown. |
| 1983 (March) | Florida (US) | Damage to thermal shield and core barrel support causing two-year shutdown for repairs. |
| 1984 (Sept.) | Alabama (US) | Safety violations and design problems force six-year shutdown for repairs. |
| 1985 (March) | Alabama (US) | Instrumentation system malfunction caused shutdown. |
| 1986 (April) | Massachusetts (US) | Emergency system fails causing immediate shutdown. |

(*Continued*)

**Table 2.9 (Continued)**    Information on Worldwide Reactor Accidents, Shutdowns, and Radioactive Leaks

| YEAR | LOCATION (NO OF REACTORS) | CAUSES OF ACCIDENTS, SHUTDOWNS, AND RADIOACTIVE LEAKS |
|---|---|---|
| 1987 (March) | Pennsylvania (US) | Cooling system fails; reactor shutdown due to overheating. |
| 1987 (Dec.) | New York (US) | Reactor shutdown due to excessive heat. |
| 1988 (Sept.) | Virginia (US) | Refueling cavity seal fails causing immediate shutdown. |
| 1989 (March) | Arizona (US) | Atmospheric dump valve fails causing emergency shutdown. |
| 1989 (March) | Maryland (US) | Cracks found in pressurized heater sleeve cause immediate shutdown of the reactor. |
| 1991 (Nov.) | New York (US) | Safety and fire problems cause immediate shutdown. |
| 1992 (April) | N. Carolina (US) | Emergency forces the reactor to shut down. |
| 1993 (Feb.) | Texas (US) | Auxiliary feed water pump fails causing reactor shutdown. |
| 1993 (March) | Tennessee (US) | Emergency reactor shutdown. |
| 1993 (Dec.) | Michigan (US) | Main turbine experienced major failure due to maintenance problems leading to immediate reactor shutdown. |
| 1995 (Jan.) | Maine (US) | Cracks found in steam generator tubes causing shutdown. |
| 1995 (May) | NJ (US) | Ventilation system failure leading to reactor shutdown. |
| 1996 (Feb.) | Connecticut (US) | Leaking valve and multiple cracks forced shutdown. |
| 1996 (March) | Florida (US) | Equipment failure causes immediate shutdown. |
| 1996 (Sept.) | Illinois (US) | Cooling water system failed causing reactor shutdown. |
| 1997 (Sept.) | Michigan (US) | Ice condensation system failed causing shutdown. |
| 1999 (May) | Connecticut (US) | Steam leak in feed water heater caused shutdown. |
| 1999 (Sept.) | New Jersey (US) | Freon leak damaged cooling system causing immediate reactor shutdown. |
| 2002 (Feb.) | Ohio (US) | Severe corrosion of control rods cause reactor shutdown for two years. |
| 2003 (Jan.) | Michigan (US) | Main transformer fire damaged the system causing immediate reactor shutdown. |
| 2005 (June) | NJ (US) | Leakage of tritium and strontium from the reactor core into the lake caused damage to main generator and turbine leading to reactor shutdown. |
| 2006 (March) | Tennessee (US) | Spill of 20 liters of highly enriched uranium from the fuel plant caused seven-month shutdown of the reactor. |
| 2009 (Nov.) | Pennsylvania (US) | Radioactive dust from the reactor contaminated 12 workers, which caused immediate shutdown. |

*(Continued)*

**Table 2.9 (Continued)**   Information on Worldwide Reactor Accidents, Shutdowns, and Radioactive Leaks

| YEAR | LOCATION (NO OF REACTORS) | CAUSES OF ACCIDENTS, SHUTDOWNS, AND RADIOACTIVE LEAKS |
|---|---|---|
| 2010 (Jan.) | New York (US) | 600,000 gallons of radioactive steam was vented out after shutdown of the reactor. |
| 2010 (Feb.) | Vermont (US) | Leakage of radioactive tritium from an underground pipe into the ground water supply caused reactor shutdown. |
| 2013 (June) | Washington State (US) | Leakage of 1 million gallons of tritium from a rusty storage tank into the underground water supply caused immediate reactor shutdown. |

*NOTE:*   The accuracy of the nuclear plant-related problems and reactor shutdowns cases as mentioned in the above tables is strictly dependent on the accuracy of the reference document. However, the author feels that the reactor shutdown decisions by the nuclear plant authorities are reasonably accurate and meet the plant and personnel requirements.

by the author reveal that a nuclear power reactor and its associated components such as circulatory cooling pumps, heat exchanger tubes, and other critical auxiliaries must receive efficient regular maintenance service to avoid fires; explosions in the core; leakage of coolant, radioactive gases, or liquids from the reactor zone; and malfunctioning of cooling auxiliaries. The most serious maintenance or repair problems in nuclear power reactors are due to lack of inadequate cooling. It should be stressed that typical operating temperatures in the core zone range from 4,500 to 5,000°C and, therefore, immediate removal of heat generated in the core must be the principal design requirement. Data presented in Table 2.8 [6] indicate that this immediate removal was not considered as the principal design requirement in some reactors.

Table 2.9 [6] includes data identifying the serious problems occurred.

## 2.11 Summary

The essential components of a nuclear power plant have been described with particular emphasis on reactor core, type of nuclear fuels, cooling elements, and system auxiliaries. Instant liberation of heat energy released from the fission process has been described briefly. Conversion of neutron energy from MeV units to conventional

CGS system has been described. Kinetic energy of thermal neutrons has been explained in great detail providing appropriate mathematical expressions. We also discussed the critical properties of rare earth elements and their isotopes produced during the fission of a reactor. Operating requirements for a nuclear reactor using various fuels, moderators, nuclear shields, and coolants have been summarized with particular emphasis on safety, reliability, and minimum risk due to radiation danger. Cost-effective tradeoff studies results using specific fuel, moderators, and cooling technology with minimum radiation are summarized in detail for the benefit of readers. Fission products generated in the reactor and their isotopes are identified with emphasis on their applications in the medical field. The critical operational status of a nuclear reactor has been explained with appropriate types of moderators and coolants. Tradeoff study results on the reactor operated over certain period and then after a specified cooling period are summarized in great detail. Rough estimates of beta and gamma energy levels released by the fission products over one year of operation are shown in Figure 2.3. The most serious reactors accidents due to cooling or overheating, which happened in various countries around the world, have been summarized identifying various problems such

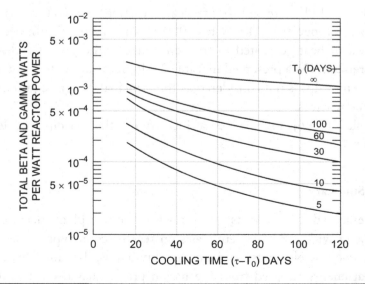

**Figure 2.3**  Total beta-particle and gamma-ray power components generated from the fission products in the nuclear reactor.

as cooling, overheating, reactor core melt-down, or reactor shutdown. It is very important that regular maintenance procedures must be followed to reduce costs.

## References

1. Samuel Glasstone, *Principles of nuclear reactor engineering*, D. Van Nostrand Co, Inc, Princeton New Jersey, July 1955, p. 85.
2. Samuel Glasstone, *Principles of nuclear reactor engineering*, D. Van Nostrand Co, Inc, Princeton New Jersey, July 1955, Table 12.2 (modified), p. 746.
3. Samuel Glasstone, *Principles of nuclear reactor engineering*, D. Van Nostrand Co, Inc, Princeton New Jersey, July 1955, Table 2.13 (modified), p. 117.
4. Samuel Glasstone, *Principles of nuclear reactor engineering*, D. Van Nostrand Co, Inc, Princeton New Jersey, July 1955, Chapter 2, p. 118.
5. A.R. Jha, Sr., "Calculated reactor power output after ONE YEAR operation and after cooling over 30 to 100 days", technical report prepared by Jha Technical Consulting Services, 12354 Charlwood street, Cerritos, CA 90703.
6. "List of nuclear power reactor accidents country by country", prepared by union of concerned scientists and students at M.I.T., (mass) 1969.

# 3

# RARE EARTH MATERIALS BEST SUITED FOR RF AND EO DEVICES AND SYSTEMS

## 3.0 Introduction

This chapter will focus on the applications of rare earth materials for radio frequency (RF) and electro-optical (EO) devices and systems. Technical articles published previously reveal that some unique rare earth elements are best suited for RF and EO devices and systems. The author will identify those rare earth elements or materials best suited for such devices and systems. In addition, some unique characteristics of such materials will be briefly mentioned that will justify their applications.

### 3.0.1 Rare Earth Doping Materials

Scientific research studies reveal that trivalent rare earth materials that are known as dopants, such as thulium (Tm), erbium (Er), or holmium (Ho) play critical roles in the design and development of EO devices and systems. Note that these dopants emit infrared radiations at 2.94 micron, 2.10 micron, and 2.04 micron, respectively. Simulated energy then extracts optical energy from the excited atom as a laser beam. In brief, these rare earth elements are ideal for the development of low-power solid state laser components operated at those specified wavelengths. It is important to mention that strong absorption lines at the desired wavelengths can be obtained by adding the right amount of rare earth dopant to the solid-state host crystal to absorb diode-pumped laser light and to transfer its energy to the dopant atom. Figure 3.1 shows a block diagram of such laser system using rare earth host crystal and diode-pump.

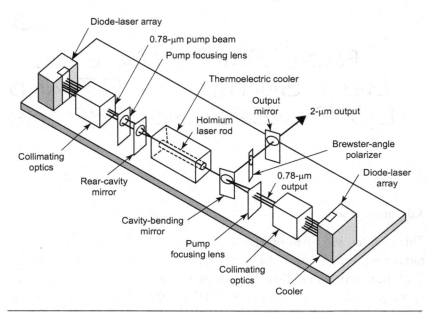

**Figure 3.1** Critical components of a solid state IR laser deploying solid state pumping technology.

### 3.0.2 Trivalent Rare Earth Dopant Materials

**Thulium (Tm):** this rare earth element offers infra red (IR) radiation at 2.10 micron and efficiently extracts the optical energy without any distortion. It has valences of 2 and 3. This material is considered as valuable as gold, silver, or cadmium, costing around $1,200 per pound. This element has close to 16 isotopes that have potential applications in commercial products. Note that its isotope Tm-169 is very stable and is best suited for commercial applications. Rare earth scientists claim that its isotope Tm-171 has potential applications in energy-producing areas. It is important to mention that Tm is widely used as a ceramic-based magnetic material in the development of electronic components. This ferrite is best suited in the design and development of acoustic-optic devices such as tunable optical oscillators, and microwave components such as isolators and circulators, which are deployed to protect sensitive radar transmitters and receivers from microwave spikes. It should be noted that this rare earth material is moderately toxic and, therefore, must be handled with extreme care.

 **Erbium (Er):** this rare earth material is a critical element widely used in the design and development of fiber optic transmission cables and fiber optic amplifiers, where superior overall performance and minimum transmission losses are of serious concern. Optical engineers believe that these fiber optic amplifiers offer an optimum as well as stable gain over wider spectral bandwidths under harsh operating conditions. Er-based optical crystals [1] are widely used in the design of dual-valence and trivalent laser systems such as Er:YLF (erbium-yttrium-lithium-fluoride) and Er:Ho:YLF (erbium-holmium-ytterbium-lithium-ferrite) rare earth–doped crystals, which are classified as dual-valence and trivalent optical linear crystals. It is interesting to point out that rare earth nonlinear crystals generally yield poor conversion efficiencies. It should be further noted that higher differential quantum efficiency, enhanced electrical-to-optical efficiency, and improved beam quality with minimum cost and complexity are possible through optimum pumping schemes. There are two different pumping schemes, namely, diode-pumping and lamp-pumping schemes. Note that tradeoff studies must be undertaken to determine which pumping scheme will offer lower cost and complexity. It is interesting to mention that the tradeoff studies undertaken by the author seem to indicate that the lamp-pumping scheme would be ideal for lasers demanding higher optical power output. However, this particular scheme would not yield minimum weight and size of the package. Research studies undertaken by the author reveal that the lamp-pumping scheme could deliver CW power output greater than 500 mW over 3.9 to 4.1 micron range when operating under cryogenic conditions [2]. Note that higher CW power output levels have been observed under the lamp-pump scheme compared to the solid state pumping scheme. In general, rare earth nonlinear crystals offer low conversion efficiency under room-temperature environment. However, significantly higher conversion efficiencies have been observed at cryogenic temperatures, which will increase the cost and package dimensions besides causing poor reliability.

**Holmium (Ho)**: research studies performed by the author seem to indicate that doping of Ho host crystal emits optical radiation at 2.05 micron, but cannot be considered an efficient optical pumped laser at room temperature. However, the use of a sensitizer material can increase optical pumping schemes. When a host crystal such as yttrium-aluminum-garnet (YAG) is co-doped with another rare earth element such as Er or Ho, significantly more efficient laser action is observed because of the energy transfer from one ion to another. It must be stated that a Tm:Ho:YAG laser emitting at 2.10 micron is best suited for eye-safe, coherent laser systems [3].

The laser systems using the rare earth material nonlinear crystals described can be classified as CW or as pulse wave (P-wave) laser systems. It should be noted that high-power laser systems are widely deployed for the detection and acquisition of military targets and missile illumination. High-power lasers can be used for commercial and industrial applications. Lasers with lower or medium power output levels are best suited for medical applications. Typical applications of lasers using rare earth materials can be summarized as follows:

- Detection, tracking and identification of airborne targets
- Commercial and industrial applications of laser such as precision cutting of metallic beams or plates
- Deployment in IR-guided missiles
- CW laser illuminator for target illumination
- Precision laser-guided bombs with precision circular error probability better than +/– 5 mrad
- Optical and dental procedures
- Precision eye examination and eye surgery
- Note that the beam stability of a laser is of critical importance for eye examination

Research studies performed by laser scientists state that the rare earth element Ho, which emits IR radiation at 2.05 micron, is not considered to be the most efficient for optical pump laser systems. Research studies further indicate that sensitizer materials can be used to increase the efficiency of the optical pumping system. It should be noted that when a host crystal such as YAG is co-doped with another rare earth element such as Er or Ho, significantly more efficient laser action has been demonstrated due to the energy transfer from one ion to another. However, when a neodymium-doped YAG is deployed, a thermal broadening effect is observed besides the appearance of a Lorentzian line in the output of the laser spectrum.

Laser systems using rare earth doping will be investigated in detail. Laser scientists indicate that the YAG crystals play an important role in the design and development of Tm:Ho:YAG laser systems

[3] emitting at 2.10 micron. It should be mentioned that these laser systems are best suited for eye-safe coherent radar applications. This means that such a laser system is ideal for eye examination or eye surgery. In addition, these laser systems are best suited for pilots, which are constantly searching, tracking, or identifying airborne targets with great accuracy. One must note that such CW laser systems with low power levels are best suited for medical applications, whereas high-power CW lasers in general are essential for military applications. Regardless of applications, beam stability and constant beam width are of critical importance if efficiency and reliability are the critical performance requirements.

In summary, it can be stated that a laser system uses the natural oscillations of atoms or molecules between energy levels to generate coherent electromagnetic (EM) radiation in the ultraviolet, visible, or infrared regions of the EM spectrum. Note many energy levels are generated in the laser output due to REE ions, according to Einstein's theory developed in 1916.

### 3.0.3 Properties and Applications of Potential Rare Earth Elements

In general, the application of rare earth materials is strictly dependent on the availability and cost of the material. The author summarizes the costs of the rare earth materials that are available in abundance. The costs of rare earth materials most widely used in commercial, industrial, and research activities and their typical applications are briefly summarized in Table 3.1.

### 3.0.4 Rare Earth Oxides and Alloys, and Their Commercial, Industrial, and Other Applications

The author will now focus on the critical properties and applications of rare earth oxides and alloys and their typical commercial, industrial, and scientific applications.

*3.0.4.1 Rare Earth Oxides*  Most popular rare earth oxides and their outstanding properties will be briefly discussed. Rare earth oxides, which are widely used in commercial, industrial, and defense applications, can be summarized as follows:

**Table 3.1**    Estimated Cost of Widely Used Rare Earth Materials and Their Typical Applications

| RARE EARTH MATERIAL | COST IN $ PER POUND | TYPICAL APPLICATIONS |
|---|---|---|
| Cerium (Ce) | 75 | fiber optic components |
| Dysprosium (Dy) | 19 | Mobile phones, computer tablets |
| Erbium (Er) | 456 | IR lasers |
| Europium (Eu) | 4,000 | Fiber optic transmission lines |
| Gadolinium (Gd) | 250 | MRI, computer tomography |
| Holmium (Ho) | 500 | IR lasers |
| Lanthanum (La) | 70 | Thin film, microwave filters, fuel cells |
| Lutetium (Lu) | 8,000 | Imaging devices, optical coatings |
| Neodymium (Nd) | 120 | Magnets for motors and generators |
| Praseodymium (Pr) | 225 | batteries, used to develop mischmetal |
| Samarian (Sm) | 250 | Magnets for electric auto components |
| Scandium (Sc) | 750 | Doping material |
| Yttrium (Y) | 200 | Microwave filters, IR lasers |
| Ytterbium (Yb) | 300 | Optical coatings, thin films |

*NOTE:*    These estimated prices per pound have been obtained in 2014 and are strictly depended on supply and demand, purity level, and availability in the free market. It should be noted that only two to four suppliers are available for these materials. It should be further noted the availability of some of these materials may be highly questionable in case of military conflicts or disagreements between prime suppliers.

*Cerium Oxide ($Ce_2O_3$)*    This particular rare earth element is available as a metal or oxide or compound. Ce oxide is widely deployed for optical coatings and thin-film applications for microwave filters operating under room temperature and cryogenic temperature. Microwave filters using thin-film technology demonstrated lower insertion loss at microwave frequencies. Its oxides are insoluble in fluorides, but soluble in chlorides, nitrates, and acetates. Ce compounds are best suited for microminiaturized devices and components, which have potential applications in commercial and defense-related systems. It is interesting to mention that its isotopes are stable and non-radioactive and are best suited for medical and clinical applications without any fear. It has about seven isotopes, which are available for research applications with minimum cost and complexity. Latest scientific research indicates that nano-crystals can be formed for research applications using cerium oxide. Latest research and development activities have demonstrated excellent catalytic properties of Ce oxide. This oxide can be deployed in manufacturing cheap fluorescent indicators.

*Dysprosium Oxide (Dy$_2$O$_3$)*  This oxide is widely used by mobile phones, smart phones, and computer-related devices. But this oxide is particularly deployed in the production of neodymium-iron-boron permanent magnets for generators and motors for hybrid electric and electric automobiles. This magnet offers excellent magnetic performance and improved mechanical strengths under elevated temperatures up to 100°C or more. Note that under-automobile hood temperatures ranging from 125 to 160°F are not uncommon.

It is important to mention that in industrial motors and generators, operating temperatures can approach 300°C. Note that in case of a homopolar motor, a Ne disc magnet provides an optimum and reliable magnetic performance. Its oxides are best suited for high-performance optical coatings and thin-film applications. Optical scientists believe that high-quality and precision thin films are possible from this oxide. Crystalline silicon, Ga AS and other rare earth material films [4] are best suited for solar cells. Characteristics of other oxides and films can be summarized briefly as follows:

- This oxide is available in powder form or in pellet configurations.
- As stated before, this oxide is best suited for optical coatings and microwave thin films, where device performance and reliability are the principal performance requirements.
- This oxide is available in soluble forms such as chlorides, nitrates, and acetates.
- This oxide is slightly toxic and, therefore, care should be taken during road and air transportation modes.
- Its room-temperature electrical conductivity is close to 57 micoohms-cm, while its thermal conductivity is roughly 10.7 W/meter-K.
- This particular oxide has seven isotopes.

*Lanthanum Oxide (La$_2$O$_3$)*  Thin films of this particular oxide play critical roles in the design and development of microwave and mm-wave frequency devices, where minimum weight, size, and package dimensions are the principal design requirements. It is of paramount importance to state that cryogenic cooling is necessary to achieve lowest insertion loss and improved device performance. Such performance parameters are only possible under low cryogenic operating temperatures. It should be stated that superior microwave device performance is only possible if the unloaded circuit "Q" is high at the operating frequency. Note that the unloaded device or circuit "Q" is a function of physical parameters, electrical conductivity of the metal, and the operating temperature or the cryogenic temperature. Calculated values of unloaded "Qs" of YBCO thin films on MgO and LaAlO$_3$ substrates as a function of cryogenic temperatures are shown in Table 3.2 [5]. It is evident from these plotted values that the quality factor ("Q") improves as the cryogenic temperature decreases. It should be noted that the cost and complexity of the package increase with the decrease in the cryogenic temperature.

Lithium oxide is best suited for photographic and optical equipment. Thanks to this oxide, these pieces of equipment may use very sophisticated lenses to achieve very high resolution. In summary, it can be stated that La-based optical glasses with a high index of refraction and low dispersion are ideal for these types of equipment, where resolution is of critical importance. Research studies performed by the author reveal that the second potential is the development of high-temperature, high-voltage, and high-permittivity ceramic capacitors, which are best suited for power electronic applications. It is important to mention that multilayer and laminated high-temperature ceramic capacitors such as BaNd$_2$Ti$_4$O$_{12}$ need optimum sintering temperature close to 1,350°C. Note that La oxide offers slightly lower sintering temperatures [6].

It is interesting to mention that in the beginning there were two types of optical glasses, namely, crown glass, which is essentially a

**Table 3.2**  Performance Comparison and System Parameters for DHSS and FPSS Lasers

| SYSTEM PARAMETER | DPSS LASER | FPSS LASER |
|---|---|---|
| Laser efficiency (%) | 10 | 12 |
| Laser efficiency with cryogenic cooling | 15 | 17 |
| Size | small | 22% larger |
| Weight | light | 30% heavier |
| Reliability (MTBF in hours) | 10,000 | 1,800 |
| Beam quality | Excellent | Slightly poor |
| Amplitude stability | High | Moderate |

soda-lime-silicate, and flint glass, which is a lead-alkali-silicate. However, by incorporating other oxides such as barium oxide or La oxide we have created a new family of glasses, namely, barium crowns, barium flints, La flints, and La crowns, which significantly improved optical properties.

It is interesting to point out that the La glasses have lower silica content, but other additives can be added to provide stability in the viscosity and crystallization temperature. La oxide is deployed in glass manufacturing because of its lower cost, namely, less than $70 per pound depending on the purity index. Note that the addition of La oxide improves the index of refraction and Abbe number, which are considered as ideal parameters for special optical glasses. The parameters as specified above yield low dispersion in glass, which is best suited for specific applications.

*Neodymium Oxide ($Nd_2O_3$)* It is important to mention that Er oxide, Pr oxide, and Nd oxide are all widely deployed in glass-coloring applications. Nd oxide is particularly deployed in the development of highly efficient optical crystals, which are best suited for lasers operating at room temperatures and emitting at specified infrared wavelengths. Note that the performance of colored glassed is dependent on glass thickness, concentration of Nd content, and source of illumination. This particular oxide is best suited for art glasses and special filters with concentration ranging from 1 to 5%.

Nd welding glasses are well known to the manufacturing steel companies, welding shops, and lamp workers. Note that such welding glasses protect the eyes of workers from the yellow flare emitted by the sodium vapors originating from hot glass sections or fluxes. This is strictly due to the narrow absorption peak of the Nd ranging from 500 to 590 nm. Note that the sodium atoms emit yellow light with a peak ranging between 500 and 590 nm.

- European oxide ($Eu_2O_3$)
- Lutetium oxide ($Lu_2O_3$)
- Praseodymium ($Pr_2O_3$)
- Samarium oxide ($Sm_2O_3$)
- Ytterbium ($Yb_2O_3$)

*3.0.4.2  Rare Earth Alloys for Permanent Magnets*

*Neodymium Cobalt (NdCo$_7$)*  NdCo permanent magnets are best suited for radar transmitters and communication RF sources operating at unattended locations. Because of their high magnetic energy capability, NdCo magnets are ideal for unattended applications. In the last decade, these permanent magnets have been widely used in the design and development of motors and generators for hybrid electric and electric automobiles. Comprehensive research studies undertaken by the author seem to indicate that these permanent magnets offer a most reliable and superior magnetic performance under severe road conditions and elevated thermal environments. These magnets have demonstrated high reliability, enhanced electrical efficiency, and improved structural integrity under rough driving conditions and operating environments.

Automobile engineers have observed a substantial decrease in the weight and size of the generators and motors when using NdCo permanent magnets. They have noticed that these magnets operate safely and reliably at elevated temperatures close to 125°C.

It is interesting to mention that Nd permanent magnets offer maximum energy product (BH$_{max}$) even at elevated temperatures. Besides their applications in automobiles, these magnets are best suited for medical diagnostic equipment such as MRI and CT, where high resolution and improved accuracy are the main requirements. While using these magnets, there is a possibility for a reduction in vibrations, irritating noise levels, and discomfort to patients.

NdCo permanent magnets will cost slightly more than neodymium-iron-boron permanent magnets because cobalt costs more than iron. Japan has a patent for the manufacturing of such permanent magnets, which is likely to expire in the near future. Note that the magnetic energy product (BH$_{max}$) for a neodymium-iron-boron permanent magnet is lower than that of an NdCo permanent magnet.

*Samarium Cobalt (SmCo) Permanent Magnets*  These permanent magnets are best suited for high-power RF sources such as high-power radar transmitters[7]. SmCo magnets have been used for high-power CW and TWTAs. Laboratory tests have demonstrated that these magnets are best suited for ECM systems. Furthermore, the RF beam was perfectly linear and parallel to the tube axis. The beam quality and focusing was found to be very impressive during laboratory tests and evaluations. Significant improvement in the efficiency, weight, Size, and power consumption were noticed by design engineers of TWTAs. Note that the reductions in weight, size, and power consumption are of critical importance, particularly when high-power TWTAs are deployed for airborne jammers. As mentioned earlier, periodic permanent magnets (PPMs) are best suited to maintain the RF beam steady and focused parallel to the helix length to maintain constant gain and low AM-to-PM characteristics.

Sm is the principal rare earth material oxide and is widely used in the production of SmCo permanent magnets. These magnets are deployed in the design and development of electric motors and generators, which are the critical components of hybrid electric and all-electric automobiles. These magnets provide optimum magnetic energy product ($BH_{max}$) and retain all magnetic properties even at operating temperatures as high as 300°C. It is not uncommon for high-power all-electric cars temperatures to approach 200°C or more under the hood. Research studies undertaken by the author seem to reveal that no other magnetic material can meet such stringent magnetic performances under harsh mechanical and thermal environments. As stated earlier, a massive use of PPM made from SmCo material in the design and development TWTAs is beneficial for critical radar and electronic warfare applications. Tradeoff studies performed by the author indicate that SmCo magnetic material has replaced the expensive neodymium-iron-boron magnets. Because of their weak spectral absorption band, SmCo magnets are used in optical filters in Nd-YAG solid state lasers to surround the laser rod in order to achieve laser efficiency by absorbing stray emissions. This rare earth compound material` in widely used in research activities dealing with separation of samples based on synergistic extraction techniques, interactions between metal ions and carbohydrates, spectroscopic studies, and chemical synthesis evaluation and characterization.

## 3.1 EO Systems and Devices

This chapter will describe the most important EO systems and devices using rare earth elements, compounds, and optical crystals. It is interesting to mention that lasers have been deployed in military applications to provide target detection, tracking, identification, and ranging. IR solid state lasers using rare earth crystals will be described with emphasis on efficiency, beam characteristics, and potential benefits in airborne system applications. Other applications of rare earth crystals such as magneto-elastic, magnetic-restrictive and EO systems will be briefly mentioned with particular emphasis on their unique benefits and potential applications.

### 3.1.1 Laser Classifications

Various types of laser using rare earth elements and rare earth–doped crystals will be briefly mentioned and their suitability for specific applications discussed. Below are the solid state laser classifications:

- Semiconductor lasers such as Ga as injection laser for ranging applications
- Rare earth–doped crystal lasers such as lithium-yttrium-fluoride (LYF) laser for space applications
- Diode-pumped solid state laser for generating infra-red power
- Lamp-pumped solid state laser for target illumination purposes
- Flash-pumped solid state laser for target detection and tracking functions
- Exceemer laser for specific military applications
- Coil (chemical-oxygen-iodine-laser) for specific military applications [4]

### 3.1.2 Diode-Pumped and Flash-Pumped Solid State Lasers Operating in the Lower IR Region

Research studies conducted by the author on these IR lasers indicate that they operate in the IR region at wavelengths ranging from 1.06 to 2.54 microns. Their conversion efficiencies range from 5 to 10%, which could be improved to close to 15% at optimum cryogenic temperatures. However, this will increase the system's cost, complexity, and maintenance. Laser performance comparison and system parameters for diode-pumped solid state (DPSS) and flashed-pump solid state (FPSS) lasers are summarized in Table 3.2 [3].

### 3.1.3 Nd:YAG Laser for Space Communication

The design and development of the Nd:YAG solid state laser program was initiated by the US Air Force in 1971 with stringent specification requirements for space communication systems. This space communication solid state laser system was developed to meet the performance requirements of an engineering feasibility model (EFM). This communication is designed to demonstrate a 40,000-km inter-satellite operation and has three channels, namely, space-to-space communication channel, space-to-aircraft communication channel, and space-to-ground communication channel. It should be noted that each communication channel has stringent performance requirements, which can be summarized as follows:

- Precision automatic acquisition and alignment by laser means are principal requirements
- Demonstration of inter-tracking accuracy of +/− 1 microradian (peak-to-peak)
- Demonstration of system operation and reliability under space environments
- Transfer of date rate better than 1,000 M bits per second
- System performance as a function of laser beam variation

*3.1.3.1 Performance Parameters of Space Communication Nd:YAG Laser* Performance parameters of space-based solid state laser systems can be stated as follows:

- Laser wavelength: 0.532 micron
- Space communication range: 40,000 km
- Angular tracking error: 1 microradian or 0.057 millidegrees
- Solid state laser output (CW): 270 mW
- Communication link range: 19,320 nm
- Data transmission rate to ground: 1,000 Mbps
- Electrical-to-optical efficiency: 1% at room temperature

*3.1.3.2 Coherent, Solid State Laser, Using InGaAsP/InP Diodes*  Research studies on solid state lasers undertaken by the author seem to indicate that diode-pumped solid states employing diode arrays and optical crystals suffer from excessive weight, low conversion efficiency, and high fabrication cost. However, a laser design approach using semiconductor laser diodes as illustrated in Figure 3.1 offers a cost-effective technique. This design approach uses InGaAsP/InP laser diodes with strained quantum well heterostructures offers CW power greater than 500 mW at 1.55 microns. It is interesting to mention that this semiconductor laser design employing strained quantum well laser diodes offers improved differential quantum efficiency (DQE) better than 45%, threshold current lower than 26 mA at room temperature, injection efficiency close to 72%, and spherical beam shape. It is interesting to mention that the full width at half-maximum (FWHM) of the perpendicular far-field pattern remains constant, but the FWHM of the parallel far-field pattern increases with the decrease in the optical power output.

This heterostructure diode employs a binary compound known as IOnP, which offers the highest thermal conductivity and heat transfer efficiency. Note that the quantum efficiency depends on the drive current, cavity length, and diode junction temperature. It is critically important to note that a cryogenic temperature close to 45 K offers improved quantum efficiency and higher optical power at 1.55 microns under lower drive current level [3].

*3.1.3.3 IR Solid State Laser Using Dual and Triple Doped Rare Earth Crystals*  Research studies undertaken by the author seem to reveal that when rare earth crystals are doped with other appropriate rare earth materials, room-temperature lasers can be achieved. For example, a triple ionized Pr-doped lithium-yttrium-fluoride (YLF) or Pr:YLF can be achieved operating at 0.64 microns or 640 nm. Similarly, a Nd-doped lithium-yttrium-fluoride (Nd:YLF) solid state IR laser is produced emitting at specified wavelengths in the IR spectrum.

*3.1.3.4 Rare Earth Crystals for Mini-Lasers*  This particular section deals with design concepts for mini-lasers and low-power solid state lasers operating at lower wave lengths in the IR spectrum. Note that the design concept of such lasers is based on ion interactions. Laser scientists believe that decreasing the number of fixed ions can improve the gain per ion for a given concentration level as demonstrated by the La oxy-sulfide ($La_2O_2S$) crystal. Published technical articles identify a new concept for the development of mini-lasers using small-size rare earth crystals containing high active ion concentrations and pumping with a monochromatic source.

It is interesting to mention that the design concept of a mini-laser was developed around 1975. The mini-laser design uses small rare earth crystals of Pr chloride ($PrCl_3$), which are excited by a laser dye and pumped by a nitrogen laser. Laser scientists believe that the most suitable material for the laser crystal is Nd-doped ultra-phosphate, for which the CW laser threshold is on the order of a few milliwatts.

Two types of laser materials are available, and they offer a high gain per unit ion. The first type of material has a high number of ions per unit length, and is rich in rare earth ions that enter as a constituent at a given concentration level. This material is used in the design of $Pr_2Cl_6$ lasers.

No other material is best suited for the neodymium-lanthanum-oxy-sulfide lasers ($PrCl_3$).

## 3.2 Rare Earth Elements for IR Detectors and Photovoltaic Detectors

Comprehensive research studies performed by the author reveal that some rare earth elements are being widely used in the design and development of photovoltaic, RF, and IR detectors. Uncooled detectors and cryogenically cooled detectors will be discussed with emphasis on sensitivity and S/N ratios.

### 3.2.1 Photovoltaic Cells

Photovoltaic cells are widely used in solar cell power modules. Note that most of the solar power installers deploy either crystalline-silicon devices or amorphous-silicon devices. Solar cell conversion efficiency is strictly dependent on the material use, cell configuration, implementation of diffusion emitters, operating temperature and humidity, photo-generation profile of the solar cell, collection efficiency, aging effect of the cells, cell junction design configuration, leakage current, and incident power density. Theoretical conversion efficiency and

**Table 3.3** Conversion Efficiency and Collection Efficiency of Solar Cells Fabricated from Rare Earth Materials

| SOLAR CELL TYPE | COLLECTION EFFICIENCY (%) | CONVERSION EFFICIENCY |
|---|---|---|
| Silicon cell | 37 | 17.05 |
| Crystalline solar cell | 39 | 18.75 |
| GaAS cell (normal) | 42 | 28.24 |
| GaAs cell (reversed) | 39 | 25.74 |

collection efficiency of solar cells using different rare earth materials are summarized in Table 3.3 [4].

It should be noted that the adverse effects of aging, temperature, current leakage, and humidity will reduce these parameters by 5 to 7%, approximately. It should be further noted that the effects of photovoltaic technology using either mono-crystalline silicon or polycrystalline silicon are not considered in the above computations. The adverse effects of component malfunction, broken cover glass, and mechanical breakdown can affect the performance, reliability, and longevity of solar cell devices. Such adverse effects are briefly summarized in Table 3.4.

## 3.3 RF and IR Detectors

RF and IR detectors using rare earth elements will be discussed briefly with emphasis on detectivity and responsivity as a function of cryogenic temperature and wavelength. Most IR detectors, quantum detectors, ternary alloy detectors such as mercury-cadmium-telluride (Hg:Cd:Te), and semiconductor GaAs detectors require cryogenic cooling to achieve significant improvement in detectivity and

**Table 3.4** Various Factors Affect the Performance, Reliability, and Longevity of Solar Modules

| COMPONENT MALFUNCTION OR MECHANICAL DEFECT | IMPACT ON PERFORMANCE, RELIABILITY, AND LONGEVITY |
|---|---|
| Broken cell, micro-crack in module | Reduces PV module performance |
| Effects of structural damage | Decreases module efficiency |
| Solder bond failure due to overheating | Affects the module reliability and longevity |
| Encapsulation discoloration due to absorption | Decreases PV module efficiency |
| Hot spots impact due to shading effects | Causes module damage and reduces efficiency |
| Corrosion due to water effect | Damages PV structural elements |

responsivity over wide spectral range, rise and fall times, dark current, noise equivalent power (NEP), and dynamic range. It is important to mention that quantum well IR (QWIR) detectors especially provide background-limited performance, and ternary compound Hg:Cd:Te detectors require cryogenic cooling well below 77 K to achieve acceptable performance for tracking at longer wavelength operations of the ICBM during the cruise phase and terminal phase of the ICBM missile.

Even Pb S, Pb Se, InAS, and Si:Ga IR detectors require cryogenic cooling around 77 K if reasonable detector performance is desired. However, for Hg:Cd:Te detectors made from ternary compound rare earth material, cryogenic cooling around 40 K or lower is essential for detecting and tracking the ICBM mission at wavelengths ranging from 14 to 18 microns.

### 3.3.1 Superconducting Detectors Operating over Wide Spectral Rages

The following superconducting detectors, as shown in Table 3.5, require cryogenic cooling [5] well below 77 K if optimum gain and lower dark current are desired:

### 3.3.2 Infrared Focal Planar Arrays

A review of recent published research articles indicates that a single cryogenically cooled infrared focal planar array (IRFLA) is capable of meeting the needs of multiple IR detectors for covert military system, space, and metrological applications. It is important to mention that the ternary material Hg:Cd:Te rare earth ternary compound is capable of detecting IR wavelengths over 1 to 14 microns IR spectral regions

**Table 3.5** Super Conducting Detectors for Optimum Gain and Low Dark Current Levels

| DETECTOR TYPE | IR SPECTRAL RANGE (MICRON) | CRYOGENIC TEMPERATURE (K) |
| --- | --- | --- |
| Ge:Hg | 8–14 | 27 |
| Si:As | 8–30 | 5 |
| Si:Ga | 8–16 | 27 |
| Hg:Cd:Te | 8–14 | 77 |
| Hg:Cd:Te | 10–18 | 38 |

at a cryogenic temperature 77 K. In brief, this ternary detector compound offers reasonably good performances in all IR bands including SWIR, MRIR<, and LWIR wavelengths. Note that Hg:Cd:Te (IRFPA) can be operated at moderate cooling temperatures, which can save considerable weight, size, and power consumption of the cryogenic cooler. Research further indicates that uncooled detectors with small junction areas suffer from large dark currents, tunneling currents, and surface generation-recombination currents. It should be stated that diffusion current level is the most dominant junction current at operating temperatures exceeding 100 K, which can seriously affect the detector's figure-of-merit (FOM). In summary, it should be mentioned that low frequency noise current, surface leakage current, and FOM are the most critical parameters of the long-wave infrared (LWIR) detector when operating in the 8–14 microns spectral region. It should be further stated that the presence of various noise mechanisms at the detector junction can significantly degrade the performance of the LWIR focal planar array.

### 3.3.3 Electro-Optical Devices

In the EO category, two EO devices will be discussed, namely, high-speed EO modulator and fiber optic amplifier.

*3.3.3.1 EO Modulator*  This particular EO device can be designed in FIVE different configurations, each using rare earth materials, namely, titanium (Ti) and lithium niobate . These modulators can be summarized and classified as follows:

(A) Ti:Li $NbO_3$ interferometer modulator operating at 0.63 microns (V/df: 1.5 v/GHz).
(B) Ti:Li $NbO_3$ interferometer modulator operating at 0.63 microns (V/df: 3.5 v/GHz)
(C) GaAS semiconductor waveguide double-heterojunction polarization modulator operating at 1.06 microns
(D) Ti:Li $NbO_3$ travelling wave modulator operating at 1.32 microns
(E) Ti:Li $NbO_3$ waveguide modulator with Y-branch design operating at 0.63 microns

Potential applications include high-speed radar signal processors, high-speed A/D converters, light-wave systems, telecommunication systems, communication systems, and monolithic-optical and optoelectronic technology integration. The most popular EO modulator is the travelling-wave modulator using a coplanar strip-line electrode on a z-cut Li NiO$_3$ substrate.

**Research and development activities must be directed towards:**

1. Techniques to reduce the drive voltage
2. Search for new EO materials with high EO coefficients ($n_1{}^3$)
3. Efficient integration of monolithic-optical and optoelectronic technologies

*3.3.3.1.1 (A) Highlights of High-Speed EO Modulators* Highlights of this EO device can be summarized as follows:

- EO modulators are the most critical components for complex radar signal processors, electronic warfare equipment, and light-wave communication systems.
- These modulators employ birefringent rare earth crystals, polarizers, and other appropriate optical components designed to operate on the polarization of a propagating electric wave signal.
- It is critically important to point out that EO modulators are based on EO effects, acousto-optic effects, or magneto-optic effects. Note the EO modulator described here is strictly based on the EO effect.
- Note that in this design concept, the EO effect changes the refractive index of the optical material as a function of applied voltage or drive voltage. However, the overall change in the refractive index includes the linear and quadratic components of the refractive index depending on the off-diagonal tensor components and the magnitude of the applied voltage.
- Open-loop and closed-loop system operations using EO modulators are possible and the component requirement is minimum for open-loop systems using travelling wave modulators.

*3.3.3.2 Fiber Optic Amplifier and Its Applications* The Er-doped fiber amplifier (EDFA) shown in Figure 3.2 is the most critical component of the communication system. Basically, there are two types

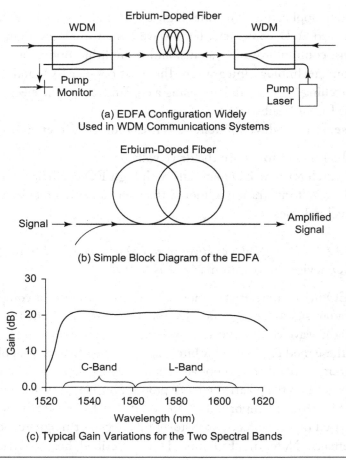

(a) EDFA Configuration Widely
Used in WDM Communications Systems

(b) Simple Block Diagram of the EDFA

(c) Typical Gain Variations for the Two Spectral Bands

**Figure 3.2**    (a) EDFA configuration, (b) amplifier block diagram, and (c) gain variation.

of optical communication systems, namely, wavelength division multiplexing (WDM) and dense-WDM (DWDM) communication systems. In case of long-distance communication systems one must consider fiber transmission loss, fiber diffraction loss, mismatch loss, and scattering loss. In order to compensate for these losses, the deployment of a fiber optic amplifier becomes a matter of necessity. In an optical amplifier, the energy is inserted into a gain medium that consists of molecules, atoms, and electron-hole pairs. The added energy excites the gain medium to a higher energy level. When a photon from the incoming signal passes through the gain medium, that photon can stimulate the emission of an identical photon, leading to the amplification of the incoming signal. It should be noted that as a solid state amplifier suffers from polarization sensitivity, coupling mismatch, and inter-modulation distortion, other techniques need to be explored. In order to boost the amplifier gain and bandwidth, other suitable techniques such as fiber optic amplifier must be explored.

Comprehensive research studies reveal that the EDFA approach is the most suitable method for boosting the gain and bandwidth of the communication systems. It should be noted that the boosted gain is due to the results of the stimulated emission of photons from an excited population of $Er^+$ ions within the Er-doped optical fiber. The usable gain of EDFA falls between 1,530 and 1,560 nm as illustrated in Figure 3.2. Note that this bandwidth of 30 nm is known as the C-band window. It should be mentioned other components such as pumping lasers, isolators, gain-flattening filters, optical couplers, and emission rejection filters will be required. The EDFA-inherent bandwidth will be affected by the parameters of the components mentioned here. In case of long-distance optical communication, system chromatic dispersion and dispersion slope must be considered. It is interesting to mention that the EDFA must be designed for longer wavelength amplifier performance. Some amplifier-inherent problems can be reduced by operating the system from the zero-dispersion wavelength which varies from 1301.5 to 1321.5 nm. Computed values of dispersion as a function of zero-dispersion wavelength and operating wavelength are summarized in Table 3.6 [6].

In case of long-distance metropolitan networks, chromatic dispersion D $(\lambda)$ can be expressed as

$$D(\lambda) = \left( S_o / 4 \right) \left[ (\lambda)_0^4 / (\lambda)^3 \right] ps / nm - km \qquad (3.1)$$

where

$S_0$ = the zero-dispersion slope whose value is less than 0.092 ps/ nm/km

ps = picosecond

nm = nanometer

km = kilometer

**Table 3.6**  Computed Values of Dispersion as a Function of Operating Wavelength and Zero-Dispersion Wavelength (ps/nm/km)

| | | ZERO-DISPERSION WAVELENGTHS | |
|---|---|---|---|
| OPERATING WAVELENGTH (NM) | | 1,310 NM | 1,321 NM |
| 1,500 | | 0.201 | 0.208 |
| 1,550 | WSX | 0.180 | 0.188 |
| 1,600 | | 0.165 | 0.171 |
| 1,650 | | 0.151 | 0.156 |

$(\lambda)_0 \lambda$ = zero-dispersion wavelength which varies from 1301.5 to 1321.5 nm

$(\lambda)$ $\lambda$   operating wavelength (nm)

### 3.3.4 Alternate Ways to Boost the Amplifier Bandwidth

Recently published technical articles indicate that there are ways to boost the amplifier bandwidth. It must be noted that a simulated Raman scattering (SRS) technique can be used to boost the optical amplifier bandwidth. Raman scattering was discovered by Indian scientist Dr. C.V. Raman. It is interesting to mention that SRS is a physical phenomenon in which a pump photon is converted into a low-frequency optical photon. When the pump power is high enough, the incoming signal stimulates the process and provides signal amplification over a wider bandwidth than that currently available from the EDFA amplifier.

Note that the rapid development of Raman amplifiers was not pursued with reasonable speed and therefore the development of the Er-doped fiber amplifier was given priority. In addition, the Raman amplifier requires higher pump-power levels. Now the fundamental problem is solved due to the availability of high-power solid state lasers. It is important to mention that expanding the total bandwidth of the fibers is necessary and requires a combination of existing technologies. Other potential rare earth dopants such as Tm or Yb besides $Er^{3+}$ must be investigated to boost the fiber bandwidth without affecting its dispersion characteristics. Laser scientists believe that the hybrid amplifier concept, which combines two technologies, offers great promise for fiber bandwidth improvement without impacting the fiber dispersion characteristics.

It should be noted that the concentration of $Er^{3+}$ ions within the fiber will determine the increase in the fiber amplifier's bandwidth. As stated before, the usable gain of the EDFA falls between 1,530 and 1,560 nm, which is known as the C-band window. This window bandwidth can be affected by the presence of other components such as isolators, couplers, gain-flattening filters, and emission filters.

### 3.3.5 Performance Capabilities of Raman Amplifiers

It is critically important to know that Raman amplifiers require sufficient input power from solid state laser sources to stimulate the Raman effect in the fiber transmission line. Early designs of Raman amplifiers employed double-clad Yb-doped fiber pumps capable of delivering output power close to 3 watts. However, the latest and improved Raman amplifiers require lower pump power levels. Applications of these continue today in short-wavelength underwater optical links to avoid expensive Er-doped fiber amplifiers. However, recent developed solid state diode laser pumps provide enough power for underwater optical links, thereby avoiding the use of expensive Er-doped fiber amplifiers for underwater applications. It should be noted that gain ripple plays a critical role in the EDFA design. The use of cascaded amplifiers in the transmission line will not only amplify the optical signals but also the gain ripple between the wavelengths.

It is important to mention that deployment of rare earth dopants such as Tm and Yb must be investigated to boost the fiber bandwidth without affecting the dispersion characteristics. Deployment of gain-equalizing filters may be effected if high-amplitude ripples appear in the amplifier band-pass region. According to EDFA designers, the gain ripple must be kept lower than 0.25 dB if optimum amplifier performance is desired. Estimated and pump wavelengths and corresponding pump power levels for three-pump and six-pump schemes are summarized in Table 3.7 [6].

### 3.3.5.1 Impact of Gain Ripple on Optical Link Performance
According to optical EDFA designers, the gain ripple in the pass band region of the amplifier should not exceed 0.25 dB between the minimum gain and maximum gain as per specifications. Gain-equalization filters with low loss are available and must be used if required to keep the ripple

**Table 3.7** Optimum Pump Wavelengths and Pump Power Levels for Various Pump Schemes

| THREE-PUMP SCHEME | | SIX-PUMP SCHEME | |
| --- | --- | --- | --- |
| WAVELENGTH (NM) | PUMP POWER (MW) | WAVELENGTH (NM) | PUMP POWER (MW) |
| 1,423 | 1,350 | 1,404 | 680 |
| 1,454 | 190 | 1,413 | 600 |
| 1,484 | 200 | 1,432 | 440 |
| ----- | ----- | 1,449 | 190 |
| ----- | ----- | 1,463 | 76 |
| ----- | ----- | 1,495 | 54 |

amplitude as low as possible. Note that if an amplifier has a maximum gain ripple of 1 dB, it will add up span after span. To avoid this situation, the main amplifier pass band ripple must be kept as low as possible. In addition, the pump source must provide more raw power to amplify the channel gain at the low power gain.

*3.3.5.2 EDFAs Operating in L- and C-Bands*  So far the performance capability of the fiber amplifiers was discussed for C-band wavelengths ranging from 1,530 to 1,560 nm. However, there is a great demand for the development of long-band or longer-band EDFAs, which covers operations from wavelengths ranging from 1,563 to 1,610 nm. In this operating band, intrinsic lower gain poses a serious design problem and requires other techniques to enlarge this particular band.

## 3.4 Summary

Rare earth materials best suited for RF and EO devices and system components have been identified. Dual- and triple-valence rare earth materials are employed for the design and development of Er:YLF and Er:Ho:YLF crystals best suited for the solid state lasers operating at IR wavelengths. These IR lasers are widely deployed in medical and special military applications. Other rare earth materials have various commercial, Medical, and industrial applications, which will be described as the summary section progresses. Research studies performed by the author reveal that neodymium-based YAG (Nd:YAG) crystals were used in the 1970s to produce high-power laser transmitters for ground-to-space and space-to-space communication links.

Rare earth elements were used in the 1940s to produce MuMETAL material, which is widely used in steel and other high-strength structural materials.

Rare earth oxides, alloys, and compounds are widely deployed in various commercial, industrial, military, and scientific research applications. La oxide, Er oxide, Dy oxide, and Pr oxide are being used in the design and development of various commercial products such as cellular phones, computer hard drives, high-capacity fuel cells and batteries, lithium batteries for multiple commercial products, and power-full permanent magnets best suited for electric and hybrid electric vehicles. Nd and Ce oxides are widely used in manufacturing glass-based products, decorating glass products, and other glass products for mass-scale commercial products. Neodymium-iron-boron and SmCo are widely used in the generators and motors needed by electric and hybrid electric vehicles. Note that these rare earth compounds offer an ultra-high magnetic energy product ($BH_{max}$), which plays a critical role in the design of motors and generators for automobiles and magnets for high-power radar transmitters and TWTAs where temperatures can reach as high as 300°C. Note that PPMs are best suited for the helix stricture in TWTAs. As stated previously, PPM magnets keep the RF beam focused as well as parallel to the TWTA axis. These magnets offer high TWTA efficiency, constant gain, improved AM-to-PM performance, and high reliability, particularly in after-burner jamming operations [7].

Potential solid state lasers using diode-pumping and flash-pumping schemes have been described using rare earth–based optical crystals. Laser classifications have been identified for various commercial, industrial, space communication, target search, tracking and identification, and medical applications with emphasis on conversion efficiency, beam performance, and beam stability. Tradeoff studies have been performed on lasers using various pumping schemes with emphasis on conversion efficiency, power output, cryogenic cooling cost, complexity, and increase in laser cost, weight, and size. A coil has been briefly described for space and military applications, where accurate target search and tracking at long infrared wavelengths ranging from 8 to 15 microns are of critical importance during the ICBM boost, cruise, and terminal phases of ICBM missiles.

The advantages of cryogenic cooling are mentioned as well as the increase in cost, weight, size, and maintenance. IR detectors and focal planar arrays (IRFPAs) need cryogenic cooling to provide high detection efficiency, low dark current, and detector gain. IRFPAs need cryogenic cooling at temperatures well below 77 K to achieve missile detection and tracking capabilities better than 90% under all operating environments. Performance capabilities and limitations of potential rare earth–based IR detectors such as Pb s, Pb Se, InAS, and Si:Ga have been summarized with emphasis on detection overall performance, dark current, dynamic range, and NEP. However, when it comes to long-range missile detection and tracking, Hg:Cd:Te ternary compound detector arrays using rare earth elements (Cd and Te) are best suited for ICBM missile detection and tracking during the boost and cruise phases of the missile, where missile plume wavelengths could vary between 10 and 16 microns. Note that Hg:Cd:Te planar detector arrays would require cryogenic cooling well below 77 K to keep the dark current low and detection and tracking probabilities high.

High-speed EO modulators and EDFAs are described with emphasis on cost, cryogenic cooling requirements, and potential applications. EO modulators are best suited for space communications and complex signal processing of forward-looking radars and side-looking radars, where signal processing is critical in target tracking and identification functions. It is important to mention that EO modulators using travelling wave modulator configurations deploy a z-cut $LiNiO_3$ substrate because it offers high EO coefficients needed for fast and complex signal processing.

The EDFA is a critical component for both long-distance WDM and DWDM communication systems. Note that in case of long-distance communication systems, transmission losses, diffraction losses, and mismatch losses can reduce the communication distance capability. An EDFA is used for the compensation of various losses and to restore the communication distance between two locations. Laser scientists believe that a hybrid concept combination of EDFA and Raman amplifier presents great promise for wide fiber bandwidth improvement without affecting the fiber dispersion characteristic of the system.

The author has carried out comprehensive research studies on other rare earth crystals that have potential scientific application. For example, piezoelectric crystals using rare earth elements such as zinc oxide (Zn O), cadmium sulfide (Cd S), and lithium gallium oxide (Li Ga $O_2$) are best suited for piezoelectric devices and sensors.

Rare earth magneto-electric crystals such as YIG, europium iron garnet (Eu I G), and terbium iron garnet (Tu I G) are widely deployed for magneto-electric device applications. It is interesting to mention that Li ferrite, also known as spinel, is best suited for magneto-acoustic device applications, where low insertion loss, high-temperature operation, and stable device performance at acoustic frequencies are the most important requirements. Note that the Li element is widely deployed in power batteries, electro-optic devices, and a variety of EO modulators to meet specific performance requirements.

# References

1. A. R. Jha, *Rare Earth Materials: Properties and Applications*", CRC Press, Taylor & Francis Group, Boca Raton, Florida 33487, 2014, pp. 260–263.
2. A. R. Jha, *Technical Report on Solid State Lasers*, Jha Technical Consulting Services, Charlwood, Cerritos, CA 90703, p.7.
3. A. R. Jha, *Rare Earth Materials and their Properties and Applications*, CRC Press, Taylor & Francis Group, Boca Raton, Florida 33487, 2014, p. 79.
4. A. R. Jha, *Solar Cell Technology and Applications*, CRC Press, Taylor & Francis Group, Boca Raton, Florida 33487, 2010, p. 263.
5. A. R. Jha, *Superconductor Technology and Applications to Microwave, Electro-Optics Electrical Machines, and Propulsion Systems*, John Wiley and Sons, New York 10158, 1998, pp. 125–127.
6. Y.j. Yu, N. Uekawa and K. Kakekawa, *Material Letters*, Elsevier Science B V, Berlin, Germany, March 2003, pp. 4088–4089.
7. A. R. Jha, *Technical Report on Sm Co PPM Magnets for High Power TWTA for Airborne Jamming*, Jha Technical Consulting Services, Cerritos, CA 90703, 1976, pp. 2–15.

# 4

# SOLID STATE RF, EO, AND MILLIMETER DEVICES INCORPORATING RARE EARTH MATERIALS

## 4.0 Introduction

This chapter is dedicated to the design and development of radio frequency (RF) and electro-optical (EO) devices incorporating rare earth materials. Solid state devices that include rare earth materials are GaAs MESFETs, high electron mobility transistors (HEMTs), GaN high-power devices, silicon diodes, Li-Ni crystals, Hg:Cd:Te long wavelength detectors, Er- and Yb-doped optical fibers, and so on. Now the author will focus on the applications of these rare earth devices in the design and development of RF, IR (infra red), and EO system components or systems such as RF amplifiers, oscillators, lasers, IR receivers, focal planar array detectors, fiber optic amplifiers, and a host of other RF and EO systems.

## 4.1 RF Components and Systems Using Rare Earth–Based Elements

RF components and systems deployed in ground, airborne, and underwater system applications will be described. We will touch briefly on the critical elements of the systems involving the use of rare earth materials. RF and IF amplifiers, RF detectors, and focal planar array detectors will be briefly discussed with emphasis on the reliability and thermal performance.

## 4.2 Infrared Detectors

Photon IR detectors such as Pb S, Si, InAs, and InSb are best suited for room-temperature applications (Table 4.1). On the other hand,

**Table 4.1**    Performance Characteristics of Photon Detectors at Cryogenic Temperatures

| DETECTOR MATERIAL | SPECTRAL RANGE (MICRON) | CRYOGENIC TEMPERATURE (K) |
|---|---|---|
| Ge:Hg | 8–14 | 27 |
| Si:As | 8–30 | 5 |
| Si:Ga | 8–16 | 27 |
| Pb:Sn:Te | 8–14 | 77 |
| Hg:Cd:Te | 8–16 | 77, 27 |

*Note:*    Hg:Cd:Te, known as the ternary compound material, is best suited for the detection of ICBM IR signature in the boost, cruise, and terminal phases of the missile. It should be noted that the cryogenic temperatures shown here are estimated values. Note that the cost of such detectors is relatively high.

Ge:Au, Ge:Cd, Ge:Zn, and Hg:Cd:Te are widely used where detectivity and responsivity are very critical parameters.

The rare earth element tellurium (Te) is chemically related to selenium and sulfur. Its extreme rarity in the Earth's crust is comparable to that of platinum due to its high atomic number. Commercially, the primary use of Te is in the manufacturing of alloys, especially steel and copper, to improve machinability.

Te is recognized as an important semiconductor material which shows great electrical conductivity in certain directions. This depends on the atomic alignment. This electrical conductivity increases slightly when exposed to light (photoconductivity). Te is widely used for semiconductor and other industry applications. It is also used in the design and development of cadmium tellurium (CdTe) solar panels for testing and evaluation at the National Renewable Energy Laboratory to determine the conversion efficiency of solar panels. Note that some of the cadmium in the CdTe compound is replaced by Zn-forming (Cd, Zn)Te compound, which is widely used in solid state X-ray detectors. It should be mentioned that Te is widely deployed by the new phase change memory chips developed by Intel Corp.

Research studies undertaken by the author on Te reveal that this rare earth element is best suited for high-performance RF detectors, particularly when the signal wavelengths range from 8 to 17 microns.

Note that Te has very high electrical conductivity in certain directions depending on the atomic alignment. It is interesting to mention that this rare earth element adopts a polymeric structure, which contains zig-zag chains of the Te atoms. Its original cost was $16 per pound around 10 years back, but due to its massive use in solar panels, metallurgy, metal industry, and a host of scientific applications, its current cost per pound is approximately $100.

Because of its massive applications in the metal industry, its physical properties must be highlighted.

The most important physical properties of Te can be summarized as follows:

- Thermal conductivity: 3.38 W/meter-k
- Young's modulus: 43 GPa
- Bulk modulus: 65 GPa
- Shear modulus: 16 GPa
- Brinell hardness: 180 MPa

This element is widely deployed in the fabrication of IR and RF focal planar array detectors and single RF detectors. When a wide-band performance is desired, cryogenic cooling is essential, which will increase the cost, complexity, weight, and size, and decrease the reliability of the detectors. The author has provided the performance of some widely used detectors.

Various types of detectors and their suitability for specific applications are briefly discussed.

Detectors come in various types and application categories, which can be summarized as follows:

- Time domain detectors
- Frequency domain detectors
- Low-power detectors
- High-power detectors
- High-speed detectors
- Photovoltaic detectors
- Quantum detectors
- Photomultiplier (PMT) Detectors

The author will identify the applications of a few detectors with a brief description as follows:

- Quantum detectors are widely used in the visible and near-IR spectral regions. Note that cryogenically cooled quantum detectors are most suitable for fiber optic–based communication systems and optical data links. Critical performance parameters of these detectors are sensitivity, detectivity, responsivity, response time, dark current, and NEP. It is critically important to mention that quantum detectors will have to be cryogenically cooled below 77 K to achieve the background-limited performance level.

- PMT detector is considered as the most efficient photon-counting device because of its high gain and low noise factor. In the nuclear imaging system, it is the PMT that creates the high image quality and reliable performance of the imaging system, which is most desirable for medical diagnosis and for night vision sensors used by the armed forces to detect objects in completely dark environments.

- Image intensifier photon detectors (IIPDs) are considered very useful for the measurement of weak signals with improved contract. It must be noted that a multistage image intensifier is generally used to amplify photons. Note that an image intensifier reduces the spatial resolution by broadening the image of a single photon. These intensifiers are widely used for monitoring pH value or calcium concentration value. Note that a potential application of this particular instrument involves the detection of bacterial contamination in food storage centers or locations.

- Focal planar array detectors (FPA detectors) are best suited for various military and commercial applications. It is interesting to point out that uncooled FPA detectors are designed with built-in CMOS read-out circuitry and intelligence networks to provide improved and instant performance, high reliability, and most sensitive $Hg:Cd:Te$ detectors. These detectors deploy ternary compound rare earth materials and are best suited for the detection and sensing of objects in space. It should be noted that uncooled IR sensors and cameras integrated with advanced FPA technology have potential applications in various commercial and defense sectors. Note that uncooled IR sensors and cameras create images of the objects in the field-of-view (FOV) in real time. It should be noted that variations in emissivity indicate surface features that may be of significant importance in the recognition and identification of objects or quality control and assurance evaluation of industrial and military products.

The FPA detector is best suited for the detection of airborne targets. The most widely used FPA detector deploys ternary rare earth materials such as Hg:Cd:Te. For target tracking and identification of space-borne targets, cryogenic cooling below 77 K is essential. This particular FPA detector is capable of precision tracking the missile plume during the boost phase, cruise phase, and terminal phase. Furthermore, this sensor is best suited for tracking the missile plume at wavelengths ranging from 8 to 16 microns, approximately.

### 4.3  RF Amplifiers Using GaAs Transistors

Now the author will describe solid state amplifiers using GaAs MESFET and GaN devices, which deploy rare earth materials. It should be mentioned that GaAs MOSFETs and MESFETs are widely used for solid amplifiers and other circuits. However, the metal semiconductor field effect transistors are widely deployed in the design and development of microwave amplifiers and lower mm-wave amplifiers with moderate power levels and efficiencies. Note that GaAs dual-gate MESFET devices have improved performance over single-gate MESET devices in terms of power output and gain.

Dual-gate MESFET devices are best suited for gain-control activities and RF switching applications where fast switching and high dynamic range switching are design requirements. These devices are best suited for mixers with conversion gain and inherent port-to-port isolation to minimize RF filter requirements. It is important to point out that monolithic mm-wave GaAs MESFET power amplifiers for 35 GHz RF seeker applications have been designed and developed [1]. This amplifier was designed and developed using the monolithic technology described by A. R. Jha (1989) [1].

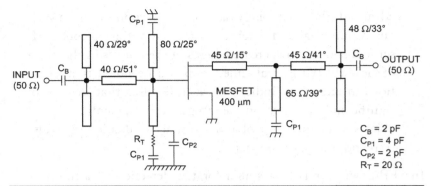

**Figure 4.1**   Circuit topology for a 35 GHz single-stage power amplifier using GaAS transistors.

This seeker GaAs MESFET amplifier was designed using a single-ended and balanced configuration on a chip including DC block, RF bypass, and bias networks. The MMIC amplifier is a cascaded four-stage amplifier which includes two-stage single-ended and two-stage balanced MMIC modules using high gain GaAS MESFET transistors. MESFET devices shown in Figure 4.1 are cascadable to provide higher gain and RF output at 35 GHz. Improvement in gain and noise temperature can be obtained by cryogenic cooling. Highlights of this amplifier can be summarized as follows:

- CW saturated power output: 300 mW
- Instantaneous RF bandwidth: 2 GHz (34 to 36 GHz)
- Channel carrier concentration: $5 \times 10^{17}$ per cm$^3$
- GaAs MESFET device parameters: 0.35 microns by 400 microns (gate length by gate width)
- Power density: 800 mW/mm
- Estimated production cost of the MMIC amplifier over the hybrid circuit technology: 33% (in 100 to 500 quantity).

### 4.4 Wide-Band, High-Power GaN Amplifiers for Radar and Electronic Counter-Counter Measures (ECCM)

This section of is exclusively dedicated to wide-band, high-power GaN amplifiers, which have potential applications in radars and ECMs. These amplifiers deploy gallium nitride (GaN) devices. It is interesting to mention that Bode-Fano theory places fundamental limits on GaN technology based on wide-band power amplifier (WBPA) design using reactive matching elements. For a typical gallium arsenide (GaAs) pseudomorphic high-electron mobility transistor (p-HEMT), the operating voltage requirement is 6 to 8 volts, while for a GaN HEMT, typical operating voltage is 30 to 40 VDC. Therefore, achievable RF bandwidth using reactive matching elements employing GaN technology is much greater than that available from GaAs technology. In addition, higher operating voltage helps in achieving higher power density.

Note that in the past decade, comprehensive research indicates that non-uniform distributed power amplifiers (NDPAs) and wide-band power amplifiers using reactive matching element technology have potential applications in the design and development of WBPAs.

*4.4.1 Application of Metal Matrix Composite (MMC) Technology for GaN Amplifiers*  Comprehensive research studies undertaken by the author on MMC technology seem to indicate that application of rare earth MMC materials such as neodymium-iron-boron (Nd:Fe:B) will realize substantial reduction in weight, size, and cost, while significantly improving the thermal performance and electrical efficiency of the RF amplifiers. Research studies further indicate that MMC technology offers advantages as summarized:

- Higher temperature capability
- Fire-resistance characteristics
- Higher transverse strength and stiffness
- No moisture-absorption capability
- Improved electrical and thermal conductivity

According to metal and mineral analysis scientists, the global MMC market size is expected to reach more than 11% by the year 2021. The experts on MMCs believe that the largest product segment is the aluminum-based metal matrix, and this is expected to account for a minimum of 30% of the demand for commercial, industrial, and aerospace applications. It is important to mention that the need for the lightweight and high-tensile strength properties of aluminum are the driving factors for the demand of aluminum-based metal matrix, particularly for aerospace applications.

Our comprehensive research studies reveal that the refractory matrix metal (RMM) composite material that contains ceramic material is expected to be the second-largest MMC technology market by the year 2025. In summary, there are two composite technologies, namely (i) MMC and (ii) aluminum-metal matrix composite. The author feels that these composite technologies can be integrated in the design and development of next generation of WBPAs involving standard HEMT and p-HEMT devices.

Details on inverted-HEMT devices and their extrinsic parameters can be expressed as a function of cryogenic temperature. Amplifier design engineers and solid state scientists claim that most research on GaN WBPAs should be undertaken with emphasis on MMIC technology compared to hybrid MIC (HMIC) technology. However, the main disadvantages of HMIC technology are the parasitic elements of wire bonding, mounting of active chips, and packaging of the assembly, which might limit amplifier bandwidth capability. Due to parasitic sensitivity, most NDPAs should be designed and fabricated using extensive evaluation procedures with minimum cost.

Recently developed fabrication materials including thermal interfacing, coefficient of thermal expansion of compliance materials, heat spreaders, various mounting processes, and integration of various RF components and packaging techniques are needed to achieve high performance and system stability.

The author plans to present a comprehensive survey of wide-band, high-power Ga N power amplifiers The performance parameters of various wide-band, high-power GaN amplifiers and their years of development are summarized in Table 4.2 [2] for the benefit of readers.

Comments on GaN power amplifier technology:

- GaN power amplifier designers feel that over the last decade, both NDPAs and WBPAs have used reactive matching elements to improve the amplifier's performance.
- GaN device scientists and amplifier design engineers feel that NDPAs offer better overall performance when the fabrication of the amplifiers deploys MMIC technology.
- Cost-effectiveness engineers feel that the high cost in the fabrication of MMIC amplifiers means that they must look for a different type of fabrication for these power amplifiers.

**Table 4.2**    Performance Parameters of Wide-Band, High-Power GaN Amplifiers

| YEAR | FREQUENCY (GHZ) | OUTPUT (W) | PAF (%) | BIAS (V) | DEVICE TECHNOLOGY/($L_G$)NM |
|------|------|------|------|------|------|
| 2000 | 1–10 | 2.25 | 20 | 15 | GaN on SiC/320 |
| 2007 | 4–18 | 4.10 | 15.6 | 35 | GaN on SiC/150 |
| 2008 | 1–50 | 0.15 | 1.55 | 15 | GaN on SiC/250 |
| 2010 | 2–18 | 14 | 28 | 30 | GaN on SiC/250 |
| 2011 | 2–20 | 15.5 | 19.40 | 30 | GaN on Si/200/DFP |
| 2011 | 2–20 | 16.15 | 26 | 30 | GaN on SiC/200/NFP |
| 2015 | 6–37 | 1.0 | 10 | 15 | GaN on SiC/100 |
| 2016 | 16–40 | 16 | 19 | 20 | GaN on SiC/150 |

**Symbol description**: GaN = gallium nitride, SiC = silicon carbide, $L_g$ = gate length, PAE = power-added efficiency, DFP = dual-field plate, NFP = no-field plate.

- GaN research scientists believe that other fabrication technologies such as HMIC technology must be explored to reduce the fabrication cost with no compromise on the amplifier reliability under severe thermal and mechanical environments.
- Most design engineers feel that HMIC technology offers tunability capability to optimize amplifier performance with minimum cost and complexity.
- Amplifier design engineers feel that the use of HMIC technology with tuning capability offers the best hope for the fabrication of power amplifiers with minimum cost.
- Amplifier designers believe that GaN dual-gate p-HEMTs offer higher gain and improved power-added efficiency.
- The author feels that cryogenic cooling will offer a significant improvement in gain, power-added efficiency, and power output. However, cryogenic cooling will increase amplifier cost, weight, and size, and reduce reliability. It should be noted that higher bias voltage and larger gate length contribute to higher output power. Note dual-gate p-HEMT devices will cost slightly more.
- In NDPAs, as the bias or supply voltage and the number of GaN transistors increase, the GaN device or cell must be loaded with a higher impedance level for optimum power level performance. This is the most useful design requirement and practically all power amplifier designers religiously follow this recommendation.

- Solid state scientists recommend that a significant research area for improving the performance of NDPAs is the use of monolithic impedance transformers.
- High-power, wide-band GaN p-HEMT power amplifiers are best suited for airborne electronic jamming systems against enemy radars and RF guided missiles. These power amplifiers are ideal for military field vehicles that perform jamming activities communication against enemy communication equipment.
- It is critically important to know that in most GaN device processors, the first field plate is formed by an overhanging T-gate structure, and the second field plate may be a source-connected or floating shield between the gate and the drain. No that these field plates improve the FOM of the transistor, which improves the power-added efficiency and CW power output of the device.

### 4.4.2 Advanced Material Technology Needed to Meet Thermal and Mechanical Requirements

The design and development of the next generation of high-power, wide-band GaN amplifiers require advanced materials capable of meeting stringent thermal and mechanical requirements. It is important to mention that the p-HEMT GaN transistor has a small structure and may not be able to dissipate heat rapidly. Note that the rapid heat dissipation theory prefers wide metallic structures with high thermal conductivity to disperse the heat as soon as possible. Microwave design engineers believe that devices dissipate more than 2 W power in hermetically sealed flange-mounted packages for rapid heat removal. Due to compact and small package requirements, design engineers must make sure that the flange-mounted package does not exceed the package specification requirements. It should be further noted that the rapid heat removal is dependent on a lower temperature environment. Note that the rate of heat removal is dependent on the effective thermal conductivity of the package material and on environment temperature. If the package contains several parts, then the effective thermal conductivity of the package should be taken into account in computing the rate of heat removal at the outside temperature. It should be mentioned that flange-mounted devices offer a large amount of heat removal to the nearby atmosphere, provided the outside temperature is moderate.

Note that the materials selected for the flange and the flange thickness must facilitate a rapid heat removal from the flange surface. As stated earlier, technologies ranging from hybrid and bare die-to-metal ceramic air cavities and to over-molded plastic surface-mount devices (SMDs) use advanced thermal interfacing materials (TIMs), heat spreading elements, and advanced epoxies with improved thermal properties and efficient thermal management technology, efficient mounting configuration, and the latest packaging technology.

### 4.4.3 Thermal Properties of Advanced Materials for the Next Generation of GaN Amplifiers

It is critically important to mention that high-power, wide-band amplifiers stability is strictly dependent on temperature-dependent material properties such as thermal conductivity, coefficient of thermal expansion (CTE), tensile strength, and elastic modulus. Note that temperature-dependent material properties are also critical in the selection of packaging materials. In order to eliminate or reduce thermal problems associated with conventional packaging materials, advanced plastic packaging materials with high power dissipation capability should be deployed. These advanced plastic materials deploy innovative, high-temperature compound plastic, epoxy, advanced TIMs, and sophisticated design techniques to retain both excellent high-power capability and significantly improved power-added efficiency. It is important to mention that the high-temperature stable performance and reliability of these amplifiers require temperature-dependent material properties such as thermal conductivity, elastic modulus, and CTE of the packaging materials. The temperature-dependent properties of some essential packaging materials, heat spreaders, and die attach are summarized in Table 4.3.

### 4.4.4 Performance Limitations of GaN Devices and High-Voltage GaN Transistor Reliability

It should be remembered that both the defense and commercial applications continue to push output power and bandwidth limitations,

**Table 4.3**  Thermal Properties of Packaging Materials, Semiconductors, Heat Spreaders, and Die Attach

| MATERIALS | THERMAL CONDUCTIVITY (W/cm°C) | CTE (ppm)/°K |
|---|---|---|
| Aluminum | 2.37 | 24.21 |
| Beryllium oxide | 2.52 | 8.25 |
| Copper | 4.24 | 17.14 |
| Alumina | 0.30 | 7.32 |
| Gold-in-tin (AuSn) | 0.57 | 16 |
| GaAs | 0.46 | 5.9 |
| GaN | 1.30 | 4.3 |
| Mg | 1.54 | 27.1 |
| Si carbide | 2.5 | 2.4 |

but GaN HEMT devices have their own limitations. Note that the gate-to-source capacitance becomes the limiting parasitic, which eventually reaches a limit where the gate periphery starts to impact on the device's operating bandwidth. It should be further noted that one straightforward method to get more output power at a given gate periphery is simply to increase the drain voltage. Both the input and output broadband-matching networks are necessary for the broadband power amplifiers. Note that the output-matching network comprises a broadband impedance-matching network, second harmonic-tuned circuits, and a third harmonic-tuned circuit to maximize the power-added efficiency of the broadband power amplifiers.

### 4.4.5  Reliability of GaN HEMT Devices

The reliability of GaN HEMT transistors is of critical importance particularly at elevated temperatures exceeding 200°C. Results of multi-temperature stress tests conducted at 65 V demonstrated a lifetime greater than 1 million hours at 200°C channel temperature. It is important to mention that, during the stress tests, the GaN transistors were biased with a source grounded at the drain voltage of 65 V, while the gate voltage was dynamically adjusted to maintain the needed drain current around 150 mA. A HEMT transistor manufacturer selected 175 transistors to evaluate MTBF from three different lots for the stress tests at five distinct test temperatures ranging from 300 to 335°C. Solid state device scientists recommend lower drain voltage, say 48 V for stress tests, in order to get reliable MTBF data at higher operating temperatures ranging from 300 to 350°C. As far as the amplifier performance at 1 dB compression point is concerned, spurious signals greater than 20 dB can be expected over more than a decade RF bandwidth. It should be noted that saturated power level beyond 1 dB compression point is acceptable for industrial applications. In some high-power, broadband power amplifiers, the output power at 1 dB compression point is very close to saturation power level. Note that am-to-pm performance in the broadband amplifiers could be unacceptable for some applications.

It is important to point out that these properties play a critical role in maintaining the performance parameters and reliability of the GaN devices at elevated temperatures approaching 300°C. The coefficient of thermal expansion will provide useful information in the selection of packaging materials.

### 4.4.6 Impact of Thermal Properties on the Reliability and Longevity of GaN Amplifiers

Research studies undertaken by the author on high-power amplifiers indicate that keeping the device junction temperature as low as possible, and stable, will provide better device performance, reliability, and longevity for a long time. The studies further indicate that proper thermal management offers a reduction in the die and overall amplifier package. Note that high-temperature stability and temperature-dependent material properties such as thermal conductivity, elastic modulus, and coefficient of thermal expansion have become absolutely critical in the selection of packaging materials.

Over the last decade, solid state scientists have performed comprehensive research studies which have reported on heat-flux cooling techniques for high-power microwave monolithic amplifiers. This particular heat-flux cooling technique is based on a thermal simulation software using the computational fluid dynamics (CFD) theory. Electronic and RF engineers believe that this simulation technique will be most beneficial for rapid heat removal from overheated RF components with ever-smaller footprints.

The everover heating condition exists in several applications, namely high-power, broadband power amplifiers, travelling wave tube amplifiers, airborne missiles, and mobile devices, where processors can heat up to more than 35°C within a few seconds, when handling computationally intense workloads. This software is designed to manage sharp heat rise to protect the high-power devices and distribute the heat effectively and efficiently thereby protecting complex and costly devices and systems.

It is interesting to mention that CFD uses the Navier–Stokes flow equations for viscous and heat-conducting fluids to model the flow of fluids through and around the objects. Note this CFD technology has been deployed over the past 25 years, it provides a useful tool for design engineers in the verification and optimization of thermal design. It must be noted that a key element in the simulation is the generation of a grid to represent a device or system as a mesh of cells. Specific details on this thermal modeling software (6 Sigma ET) are available in the technical article published in *Electronic Products* dated April 2015.

### 4.4.7 Ideal Materials for Packaging and Die Attach

As stated earlier, advanced TIMs, along with high-temperature plastics, must be given serious consideration during the amplifier fabrication phase. Both device design engineers and solid state scientists suggest that TIMs must be deployed to attach devices to the substrate or heat sink. Note that among TIM materials, solder has very low thermal resistance stress bonds but suffers from a high coefficient of thermal expansion. In addition, they result in very high thermally induced stress modulus between the device and the solder. As a matter of fact, this is especially true when solders are used to directly bond semiconductor devices with a low coefficient of thermal expansion such as GaAs, GaN, or silicon onto a metal heat spreader, which causes a break due to poor mechanical strength. To overcome this particular problem, alloys with a low coefficient of thermal expansion, such as copper tungsten (CuW), copper molybdenum copper (CuW Cu), or molybdenum copper (MoCu) must be used as carriers for die attachment. Note that these alloys are costly, heavy, and thermally less conductive.

### 4.4.8 Die-Attach Materials

Research studies performed by the author on die-attach materials seem to indicate that there are not too many materials available for this

**Table 4.4**  Most Suitable Materials for Semiconductor, Die Attach, and Heat Spreader (Arbitrary Units to Indicate Thermal Conductivity)

| MATERIALS | ARBITRARY UNITS FOR THERMAL CONDUCTIVITY |
|---|---|
| Diamond | 1,200 |
| Copper | 400 |
| Aluminum | 245 |
| Copper-tungsten | 223 |
| Molybdenum-copper | 175 |
| Gallium nitride | 150 |
| Gold-tin solder | 83 |
| GaAs | 89 |
| Gold-germanium solder | 54 |
| Kovar | 27 |

particular application. In the ideal die-attach material, the thermal conductivity parameter must be moderately high and the procurement cost must be affordable. Research studies seem to point towards silver (Ag)-filled epoxy, eutectic, and recently developed sintered epoxy, as shown in Table 4.4 [3].

### 4.4.9  Mechanical Properties of Structural Materials Widely Used in High-Power Microwave Systems

The author now focuses on the materials widely deployed in the fabrication of base plates and packaging and on their properties. Data will be presented on modulus of elasticity and specific stiffness, which is defined as modulus of elasticity divided by density of the material

**Table 4.5**  Mechanical Properties of Pure Metals and Nonferrous Metals at Room Temperature

| METAL | MODULUS OF ELASTICITY (PSI $\times 10^6$) | SPECIFIC STIFFNESS/DENSITY $\times 10^6$ |
|---|---|---|
| Tungsten | 59 | 84 |
| Beryllium | 44 | 666 |
| Zirconium | 14 | 59 |
| Gold | 12 | 17 |
| Silver | 11 | 29 |
| Aluminum | 10.3 | 105 |
| Magnesium | 6.5 | 102 |
| Tin | 6 | 25 |
| Lead | 2 | 5 |

**Table 4.6**   Figure of Merit for Microwave Faraday Rotators Using Rare Earth Oxides

| FERRITE MATERIAL | FREQUENCY (MHZ) | ROTATOR DIAMETER (INCH) | FOM |
|---|---|---|---|
| Ni-Zn | 9,000 | 0.25 | 11 |
| Ni-Zn | 11,200 | 0.15 | 110 |
| Mg | 9,000 | 0.32 | 55 |

multiplied by $10^{-6}$ in. Mechanical properties of nonferrous metals and pure metals are presented in Table 4.5.

### 4.5 Microwave Ferrites and Their Applications in Commercial and Military Fields

There are several ferrite materials which contain rare earth elements and their oxides. It should be noted that oxides of iron ($Fe_2 O_{3)}$, manganese oxide ($Mn O_3$), magnesium oxide ($Mg O$), zinc oxide ($Zn O$) and lithium oxide ($Li_2 O$) are best suited to microwave applications such as ferrite isolator and ferrite circulators. Some microwave devices such as Faraday rotators require special materials, namely nickel-zinc (Ni-Zn) or Mg. FOM and rotator diameters for microwave rotators operating at higher microwave frequencies are summarized in Table 4.6.

Studies performed by the author on ferrite components indicate that the FOM can be given as,

$$FOM = \omega T \qquad (4.1)$$

$$T = [1 / \gamma \pi \Delta H] \qquad (4.2)$$

where, w = angular frequency = $2 \pi$ f, where f is the microwave frequency (MHz), T is the relaxation time in seconds, $\gamma$ is a gyro-magnetic ratio with a value of 2.8, and ($\Delta H$) resonance line width with a value of 50 oersteds for low insertion loss. Inserting these parameters in Equation 4.1, one gets a maximum FOM equal to [$14 f_{GHz}$]. Now, one can design a rotator for microwave frequency below 14 GHz. This means that for a microwave rotator with FOM = 13 and insertion loss of 0.5 dB, the low frequency limit will be around 1,000 MHz or 1 GHz device, whether one wants to work with magnetic fields above or below the resonance. Note that the insertion loss for each of the Faraday rotators is 0.5 dB. It is interesting to mention that the best values of FOM can be obtained from the rare earth–based ferrite materials with resonance line width in the vicinity of 50 oersteds. The author feels that the FOM values increase with the frequency. It is interesting to mention that FOM can be calculated using the mathematical expressions shown for other microwave devices such as isolator, circulator, or microwave filters. Note specific ferrite materials are available for microwave devices to meet their performance specifications.

### 4.5.1 History of Ferrite Deployment for Various Applications

Comprehensive studies undertaken by the author on microwave ferrites show that these have been deployed in microwave devices and systems as early as 1956. It is important to mention that in the 1950s and 1650s several technical papers were presented on ferrites in microwave theory and techniques conferences. Six invited papers were presented on the physical properties of ferrites with emphasis on measuring techniques. The final fifteen technical papers that were presented dealt with the theory on microwave devices. Ten tutorial review papers were selected to provide elementary information on microwave ferrite applications and devices or systems widely used in commercial and military systems.

Applications of ferrite devices and systems were given serious consideration for the duration of the Cold War, from 1954 to 1977 approximately. The design and development of microwave components such as microwave filters, limiters, isolators, circulators, and others was given priority, particularly during the Vietnam conflict. Accelerated development programs on some ferrite components were initiated during the 1970s and 1980s for deployment in communication systems, radar transmitters, and electronic jamming systems. Design and development of compact and state-of-the art ferrite components such as circulators, isolators, and band-pass filters for phased array antennas and samarium cobalt PPMs for high-power TWTAs were a priority. Note these PPM magnets are widely used by high-power, wide-band TWTAs for jamming enemy radars, missiles, and ECCMs. It should be noted that these samarium cobalt PPMs can withstand operating temperatures as high as 300°C, which is not uncommon when the fighter aircraft is operating in after-burner mode. It is important to mention that

PPM magnets offer improved reliability and longevity under harsh thermal and mechanical environments.

### 4.5.2 Widely Deployed RF Ferrite Components in the Aerospace Industry

The majority of the microwave ferrite components were designed and developed around the 1960–1990 timeframe to meet the specific performance requirements of radar systems, phased array antennas, communication equipment, dual-mode TWTAs for electronic warfare equipment, satellite systems, and airborne radar-guided missiles. Note that state-of-the art microwave components with stringent reliability and longevity requirements were developed after 1990, approximately, using advanced semiconductor technology, the latest fabrication materials, surface mount techniques, mm-wave ferrite components, latest epoxy materials, and advanced packaging materials with emphasis on high strength and efficient thermal dissipation capability. The author wishes to highlight the following unique microwave, mm-wave, solid state devices, and ferrite devices, which use rare earth elements, oxides, or compounds and are widely employed in radars, communication equipment, satellite communication systems, and phased array antennas:

- Microwave ferrites are widely used in phase shifters, ferrite-tunable filters using YIG spheres, isolators and circulators, microwave filters with specific performance requirements in the pass band and band reject regions, ferrite directional couplers, and ferrite-tuned resonant cavities.
- GaAs transistors are used for mm-wave MMIC amplifiers for airborne target detection and identification, GaN HEMT devices are used for high-power, wide-band power amplifiers for active jamming and silicon and GaAs varactor diodes are used for frequency multipliers as local oscillators needed for RF mixers.

## 4.6 High-Temperature Ceramic Capacitors Using Rare Earth Materials and Their Applications

### 4.6.1 PLZT Capacitors

High-temperature ceramic capacitors are developed using a rare earth compound material known as PLZT, which is defined as "lead lanthanum zirconate titanate". This material is especially suited for high-power, high-temperature capacitors, which are critical elements of power electronics. Note all these four elements are rare earth materials. These capacitors are defined as innovative DC link capacitors for frequency converters. Note that the PLAZ link capacitor covers a capacitor range from 1 micro-farad to 100 micro-farads at a rated voltage of 400 V DC.

Another CeraLink capacitor has a rated voltage of 800 V DC and a capacitance of 5 micro-farads, which is RoHS (restriction of hazardous substances) compatible. This high-temperature, high-power capacitors permit storage of electrostatic energy in dielectrics with minimum cost and complexity.

Generation of higher energy density is possible with this ceramic capacitor technology, because this rare earth material–based ceramic compound has high permittivity, which is known as the ratio of polarization to E-field, high electric field (E), and ultra-low material loss. These ceramic capacitors deploy thin films of rare earth materials aluminum or tantalum and come with rated voltage from 10 V DC to 100,000 V DC. They are best suited for power electronics applications.

*4.6.2 Ni-Cofired Niobium Ceramics*   These ceramic capacitors deploy multilayer fabrication technology and are best suited for actuator applications because of their compact size with smaller dimensions than PZT capacitors. Note that the Nil co-fired ($Na_{0.5}$ $K_{0.5}$) $NbO_3$ capacitor has small thickness and its dielectric constant is roughly 400 compared to 2,000 for PDT capacitors, and therefore it requires less drive voltage. This capacitor is generally referred to as NKN device. Note the first "N" stands for sodium, while the second "N" stands for $NbO_3$ (niobium oxide). But the middle letter "K" stands for potassium. The mean time between failures for this particular device is very long and the current leakage is less than 15 micro-amperes. Dielectric processing is complex in the development of rare earth element–based capacitors. Important following observations must be observed during alkali-perovskite dielectric processing, according to the high-temperature ceramic capacitors:

1. Higher densification is possible at lower temperatures.
2. Devices with low resistivity require air cofiring.

3. P-type conductivity is preferred for high-power, high-voltage capacitors.
4. Niis compatible with cofiring.
5. Silicon module capacitors are best suited for applications where operating temperatures fluctuate from 240 to 280°C. The reliability of this capacitor is appropriate for various applications.

### 4.6.3 Base-Metal Cofired (K, Na) NbO₃-Based Material

This new non-lead piezoelectric material is best suited for multi-layer piezoelectric actuators. It is formed when mixed with calcium zirconate ($CaZrO_3$). This material with Ni electrodes has wide applications for multilayer piezo-actuator applications. It is important to mention that a $BaTio_3$-based multilayer material is widely used to make small and cheap magnetic field sensors, which are best suited for the measurement of magnetic field.

The advantages of Ni electrodes can be summarized as follows:

• Cost effectiveness
• High electro-migration resistance capability
• These Ni electrodes are known as ferromagnetic materials.

Despite having so many advantages, Ni electrodes also have a major disadvantage: they must be fired in a reducing atmosphere. It is important to understand the effect of firing atmosphere upon piezoelectric properties.

### 4.7 Spark Plasma Sintering (SPS) Combined with Heat Treatment to Prepare Laminated Ceramics Using Rare Earth Elements and Their Potential Applications

It is critically important to understand that laminated ceramics with different compositions require different sintering temperatures. Although the optimum sintering temperature of barium-neodymium-titanates ($Ba\ Nd_2\ Ti_4\ O_4$) ceramics is much higher than that of bismuth titanates ($Bi_4\ Ti_3\ O_{12}$) known as (BIT) ceramics, sandwiched (BNT/BIT/BNT) composite ceramics can be successfully developed with this new method. The results obtained by sintering scientists using scanning electron microscopy (SEM) and electron probe microanalysis demonstrated that the BNT layers and the BIT layer were well bonded and no significant diffusion between the layers was observed. The results further demonstrated that the temperature coefficient of the dielectric constant of the laminated ceramics was found to be much smaller than that of the BNT ceramic material.

Ceramic scientists are very impressed with the latest test results obtained on thin layers of laminated ceramics, particularly on the mechanical and electrical properties of the capacitors. Both these properties can be tailored by adjusting the thickness and material compositions of the different layers. Ceramic scientists believe that laminated ceramic capacitors, compact, thin layers of laminated ceramics, can be achieved using different ways, including tape casting, sequential slip casting, the electrophoretic deposition method, and colloidal techniques. Each ceramic device is sintered at a suitable temperature for best results. However, ceramic scientists further believe that it is not easy to sinter such compact devices sufficiently, because the material composition of each layer requires different optimal sintering temperatures.

While testing laminated ceramic, a ferroelectric $Bi_4 Ti_3 O_{12}$ (BIT) device with a positive coefficient was used to modify the negative dielectric constant of the dielectric ceramic $Ba Nd_2 Ti_4 O_{12}$, which is called (BNT) ceramic layer. It should be noted that the optimal sintering temperature of the BNT compound ceramic is around 1350°C, while the BIC ceramic layer is around 1,100°C. Based on the experimental data obtained by chemical scientists, the sintering temperature for BNT ceramic was found to be around 1,350°C and that for BIT ceramic temperature was close to 1100°C. As far as the density of the ceramic materials is concerned, the density for BNT ceramic was found to be close to 5.65 grams/$cm^3$, while for BIT ceramic it was close to 7.7 grams/$cm^3$.

Plasma scientists believe that SPS temperature can sinter materials to high density at low temperatures for short times, typically less than 10 minutes [4]. Scientists feel that the rapid sintering method has been successfully deployed to prepare inter-metallic compounds, nano-structured ceramic materials, non-equilibrium ceramics, transparent ceramics, and other materials that are difficult to sinter by using conventional sintering methods.

A new method of SPS combined with heat treatment was deployed to investigate the preparation of sandwiched (BNT/BIT/BNT) composite ceramics. This method requires three distinct steps as described below:

STEP 1: $(Ba\ Nd_2\ Ti_4\ O_{12}/Bi_4\ Ti_3\ O_{12}/Ba\ Nd_2\ Ti_4\ O_{12})$ = (BNT/BIT/BNT), which is a composite ceramic.

 In this step, BNT ceramic was prepared using the conventional sintering method.

STEP 2: in this step, calcined and pre-pressed BIT powders were sandwiched between BNT ceramic pellets and spark plasma sintered at 900°C for 10 minutes to synthesize (BNT/BIT/BNT) composite ceramic specimens.

STEP 3: in this last step, a heat treatment was used to re-oxidize the partially reduced (BNT/BIT/BNT) composite ceramic sample or specimen. Later on, the microstructures and dielectric characteristics of the above composite ceramic were investigated.

### 4.8  Solid State RF Amplifiers for Specific Military Applications

Solid state wide-band amplifiers with moderate power capability are widely used for missile tracking applications, missile seeker applications, intercept receiver applications, and compressive receiver applications. In complicated or sophisticated military receivers, an increasing number of signals per unit needs to be intercepted, sorted, and classified. Instantaneous frequency measurement (IFM) receivers are deployed to measure the frequency parameter of the incoming signal. Note that IFM receivers can intercept one signal at a time. Multi-signal tracking receivers can accurately and simultaneously intercept several time—coincident signals without loss of information on other signal parameters.

Most signal parameters such as frequency, amplitude, time of arrival (TOA), direction of arrival (DOA), pulse width, modulation type, transmission frame characteristics, probability of intercept (POI), and transmission encoding characteristics can be determined by intercept signals. Most of the information present in the signal can be obtained in a compressive receiver or in a channelized receiver.

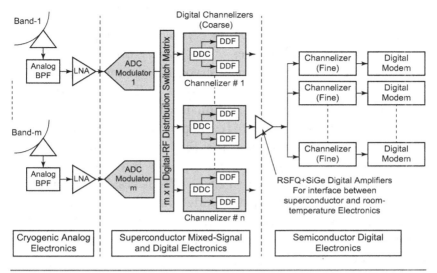

**Figure 4.2** Block diagram of multiband, multichannel, digital RF channelizing receiver configuration showing mixed-signal, digital, and superconducting electronics.

The architecture of a compressive receiver is shown in Figure 4.2, which contains important elements such as a clock generator, chirp generator, threshold detector, marker oscillator, 7.0–10.0 Ghz LO, YBCO chip filter, power amplifier, envelop detector, buffer register, and others.

## 4.9 Summary

Solid state RF and EO devices incorporating rare earth materials have been described with emphasis on performance improvement due to deployment of rare earth elements, oxides, and compounds. RF transistors such as GaAs, HEMT, MESFET, MOSFET, and GaN are described in great detail and their potential commercial and military applications investigated. Li-niobate and Te crystals are described with emphasis on their commercial and defense applications. Semiconductor transistors and diodes silicon (Si), Germanium (Ge), GaAs, and In Sb devices are described. The potential advantages of heterostructure technology are briefly mentioned in the development of high-power RF transistors such as HEMT and HBT transistors.

RF and EO systems and their components, such as RF and IF amplifiers, RF and IR detectors, and FPA detectors for detection of long-wave signals ranging from 8 to 16 microns are briefly described with emphasis on detectivity and responsivity capabilities. Photon detectors and their advantages on cryogenic cooling are summarized. The potential advantages of the ternary compound material Hg:Cd:Te are summarized under cryogenic cooling temperatures. The potential advantages and applications of Te, such as the cadmium tellurium solar panel, are summarized. It should be mentioned that the zinc-forming (Cd, Zn) Te compound is widely used in solid state X-ray detectors. Note that the Te element is widely deployed in memory chips that are being developed by Intel Corp. Research scientists feel that the rare earth element Te is best suited for high-performance RF detectors, particularly when the signal strength is close to noise level. Because of the great demand for various commercial and scientific applications, its original 10-year price per pound shot up from $10 to $100.

Various rare earth elements are widely used by RF and IR detectors. Photomultiplier detectors, Hg:Cd:Te detectors, and quantum detectors performance is significantly improved under cryogenic cooling conditions, particularly response time, dark current, and NEP. PMT detectors are widely used in night vision sensors for the armed forces. These sensors help soldiers to see objects in almost complete darkness.

FPA detectors are best suited for the detection and surveillance of targets of interest with high signal-to-noise ratio. Of course, the cryogenic cooling of these detectors is absolutely necessary to obtain high signal-to-noise ratio. FPA detectors deploy Hg:Cd:Te detecting elements, which can detect the targets present in 8 to 18 microns spectral regions. It should be noted that a high FPA detector performance comes with higher weight and cost penalty using cryogenic cooling technology. Engineers are requested to undertake tradeoff studies before using cryogenic cooling technology. FLA detector technology is best suited for the detection of plume signals of intercontinental ballistic signals during the boost phase, cruise phase, and terminal phase, whose wavelengths vary from 8 to 17 microns, approximately.

GaAs MESFET devices can be used in the design and development of 35 GHz MMIC amplifiers for RF seeker applications. Their weight and size are best suited for RF seekers, where limited package size and weight are the principal design requirements. Note that MMIC design technology has been recommended for wide-band, high-power, GaAs or GaN power amplifiers. These power amplifiers are widely deployed in airborne jamming equipment, and are best suited for aircraft location where weight, size, and power consumption are the main design specification requirements. Note that MMIC design technology meets these design specification requirements.

There is a great demand for high-power, wide-band GaN power amplifiers for airborne jamming equipment and for ECCMs. It should be mentioned that current high-performance jet aircrafts require ultra-high jamming power only for a few minutes (less than 5 minutes) when the fight-bomber aircraft is operating in after-burner mode. This means that wide-band, high-power, GaN power amplifiers will be the work-horse for the next generation of fighter-bombers. Note that power amplifiers must be limited to two-decade bandwidth capability to retain constant gain, optimum power level, and acceptable am-to-pm performance. Note that a slight drop in power in the band-pass regions of the amplifier even for a short while could degrade the effectiveness of the jamming capability. This could be dangerous, particularly when the fighter aircraft is operating in after-burner mode and insufficient power density causes an inability to provide effective jamming when needed. Deployment of the metal matrix composite technology as described by the author is best suited for wide-band, high-power designs for future applications, where high reliability and safety are the principal requirements.

According to metal and mineral scientists, the global MMCs market size is expected to grow by 12% by the year 2012. These scientists further predict that the RRMs composite material ceramic as the base material and its market will exceed 21% by the year 2025. In summary, it can be stated that three composite technologies, namely MMC, aluminum-metal matrix composite, and metal-ceramic matrix composite will play a critical role in the design and development of enclosures for housing amplifiers, oscillators, receivers, and power supply modules for the protection of aerospace modules.

As mentioned previously, HEMT devices with T-gate structures and field plates will offer higher gain, improved power-added efficiency, and significantly higher CW power output. Appropriate flange material with higher thermal conductivity and optimal flange thickness, advance interfacing materials, heat-spreading elements, advanced epoxies with improved thermal conductivity, and efficient thermal management technology will aid in the rapid removal of heat from the transistor's junctions. To maintain high operational reliability, GaN transistors must be biased with source grounded at the drain voltage around 65 V, while the gate voltage is dynamically adjusted to maintain the drain current close to 150 mA. Thermal conductivity and coefficient of thermal expansion parameters for advanced packaging materials for GaN transistors (HEMTs) are summarized. The materials suggested will help the transistors to retain their power output and reliability even when the junction temperature approaches 300°C. The SPS concept is best suited to the design and development of sandwiched (BNT/BIT/BNT) composite ceramic capacitors for high-power electronic systems, where the surface temperatures can exceed 300°C in a matter of a few seconds. Note the BNT/BIN/BNT composite ceramic capacitors use rare earth elements and their oxides such as barium, N d, Ti oxide for BNT ceramic capacitors, and bismuth, titanate, titanium oxide for the BIN ceramic capacitor. The combination of these three ceramic capacitors will lead to the formation of a laminated ceramic capacitor with outstanding parameters, which will be best suited for power electronic systems, where operating temperatures can approach 300°C. These composite ceramic capacitors will play a vital role in the design and development of high-power electronic systems.

Latest solid state GaN HEMT devices have demonstrated high-power capability at 35 GHz. It should be noted that GaN HEMT devices with gate dimensions of 0.25 microns and 200 microns and fabricated on silicon carbide substrate have demonstrated a power-added efficiency greater than 38% at a frequency of 30 GHz. Preliminary computations indicate that both the high power-added efficiency and high output power capability are not possible at the same time, regardless of trans-conductance or bias voltage.

CFD simulation based on the fluid dynamics theory offers the most desirable solution for heat removal from the solid state device junction. This simulation indicates that the heat removal capability undergoes rapid deterioration as the junction temperature increases. One must note that Navier–Stokes equations, which are a part of CFD, offer a solution to the heat removal issue as a function of junction temperature. It is important to mention that the simulation is the generation of a grid to represent a heat source. Configurations for channelized and compressed receivers, which are widely deployed by defense companies, are briefly described.

# References

1. A. R. Jha, *Monolithic Microwave Circuit Technology*, (MMIC), Artech House, 685 Canton House, Inc Norwood, MA 02062, 1989, pp. 1–28 (author of Chapter One).
2. David W. Runton et al., "History of Ga N in IEEE Microwave Magazine", IMS Special Issue, May 2013, pp. 82–92.
3. Kamal K. Samanta, "PA thermal management and packaging", *IEEE Microwave Magazine*, November 2016, p. 75.
4. Yong Yu et al., *Materials Letters*, Elsevier Science BV, 23 March 2003, pp. 4088–4094.

# 5

# USE OF RARE EARTH MATERIALS IN ULTRA-BROADBAND MICROWAVE AND MM-WAVE RECEIVERS

This chapter is dedicated to the use of rare earth materials in microwave and mm-wave receivers in the fields of communications and entertainment, and in various branches of the defense department. It should be noted that state-of-the-art non-rare earth or conventional components are readily available with minimum cost, and can be used in the design and development of entertainment equipment. It is critically important to mention that the cost of rare earth materials is very high, as shown in Table 5.1 and, therefore, their deployment in entertainment devices and equipment could not be justified.

The use of rare earth materials and devices for covert communication receivers; missile tracking sensors; wide-band signal detection; and analysis, surveillance, and reconnaissance receivers can be justified because of the need to meet stringent performance requirements. Comprehensive research studies undertaken by the author seem to indicate that intercepting several time-coincident electronic signals simultaneously may be made possible by two types of receivers, namely by compressive or channelized receivers.

## 5.0 Compressive Receiver Technology

Note that in a compressive receiver, a frequency-to-time transformation is performed and the output signals are available in serial format. This particular receiver technology is less complex and cheaper compared to channelized receiver technology. Channelized receivers are designed to meet the specific requirements for communications and

**Table 5.1**    Comparison of POO for Short Signals Bandwidth (<200 ns)

| PROBABILITY OF POO | CHANNELIZER RECEIVER | CRYSTAL VIDEO RECEIVER | IFM RECEIVER |
|---|---|---|---|
| POO (PPS = $10^6$) | <5% | 53% | 64% |
| POO (PPS = $10^5$) | <0.2% | .08% | 0.1% |

*Note:* PPS stands for pulses per second and POO stands for probability of overlap.

radar applications with particular emphasis on frequency resolution and dynamic range [1].

## 5.1  Frequency-to-Time, Domain Formation

In a channelized receiver, a frequency-to-time formation is performed and the output can be seen in serial format. Channelized receiver configuration can be described in various forms. The author will focus on parallel signal processing techniques, architectural configurations, and appropriate hardware technologies for channelized receiver configurations, which can be implemented in a compact volume with a high dynamic rage capability greater than 50 dB. There are various channelized receiver technologies such as magnetostatic wave (MSW), surface acoustic wave (SAW), bulk acoustic wave (BAW), and acousto-optic (AO) channelizer technologies. The author describes each of these technologies in great detail later on under the appropriate headings. In addition, an overview and technology trade-off study results on various receiver types will be presented for the benefit of the readers and prospective receiver design engineers.

## 5.2  Digital Signal Processing

Digital signal processing is the most important and difficult task to extract available information from the receiver. As stated earlier, the output signals from a channelized receiver are in a parallel format and, therefore, signal processing can be relatively simple. Comprehensive research studies undertaken by the author indicate that there are several methods available when it comes to parallel signal processing techniques. These can be briefly summarized as follows:

- SAW technology
- MSW technology

- BAW technology
- AO technology

Note that each of above-mentioned channelizer technology requires specific architecture. The receiver cost and complexity for each technology differ from each other. It is interesting to mention that the cost of mm-wave channelized receivers is higher than that of microwave channelized receivers. Note in parallel channelized receivers, the full electromagnetic frequency spectrum is available in several contiguous segments. Furthermore, each parallel signal processing channel of the receiver is responsible only for a specific channel band of the full segmented frequency bandwidth. It should be noted that parallel channels are independent and thus several time-coincident signals can be intercepted simultaneously without loss of information such as pulse width, frequency, and rise-or-fall time. The number of channels required for one-dimensional spectrum analysis may be as low as 25 or as high as 10,000 depending on the targets and application. Note that the cost and complexity of the channelized receiver increase as the number of channels increases.

In most cases, it is desirable to obtain both the frequency and DOA information of both coarse and fine signal frequency with electromagnetic receivers. In case of pulsed signals, additional information may be required such as pulse width, rise time, fall time, modulation type, and pulse repetition frequency (PRF). Furthermore, the complexity will significantly increase if detection of pulse compression signals is required. In cases when the receiver is facing pulse compression jamming signals, accurate detection or estimation of signal parameters will be more complex and cumbersome. In general, typical signal parameters detected by the receivers include signal frequency, amplitude, TOA, DOA, pulse width, modulation type, transmission coding characteristics, transmission frame characteristics, and POI. State-of-the-art channelized receivers are designed and developed for the detection of high-density RF signals with a high POI, simultaneous interception of time-coincident signals parameters, bandwidth, signal processing capability, and dynamic range. Other physical parameters of the receive package such as weight, size, and power consumption are of critical importance for airborne applications. It should be stated categorically that a channelized receiver would satisfy most performance requirements of RF, microwave, mm-wave for communications, radars, ECM, and ECCM applications.

### 5.3 Comparison of Probability of Overlap (POO) of Short Pulse Signals

The comparison of the POO of short pulse signals can be summarized as a function of the number of pulses per second processed during the intercept as a function of pulse rate.

### 5.4 Channelization Overview

Receiver studies performed by the author indicate that substantially greater signal throughout can be achieved with channelization signal processing techniques that have highly parallel architectures than with the conventional digital signal processing capabilities. The signal processing techniques of channelizers include spectrum analysis, precise determination of DOA of signals or direction finding (DF) sensor, correlation, excision of narrow-band signals from wide-band backgrounds (known as clutter from the background), programmable analog filtering, and computational tasks. It is critically important to point out that applications discussed here are limited to spectrum analysis, which is considered the fundamental building block for parallel signal processing.

#### 5.4.1 Dynamic Range and Speed of Spectrum Analysis

Dynamic range and speedy spectrum analyses are the main requirements for the channelized receivers dealing with advanced sensor technology such as sophisticated image sensing. This means that receiver engineers will face demanding requirements for future high-performance channelized receivers and critical channelized receiver components. Because of these stringent requirements, dynamic range and speedy spectrum analysis requirements will be upgraded. In addition, new receiver architecture will be required with full electromagnetic frequency range of interest subdivided into several contiguous channel bands. Note that the range will be separated into contiguous segments such that the full frequency range is further divided into enough parts to satisfy the desired frequency resolution. Time-coincident signals are simultaneously incident at the receiver. After going through the demultiplex process at the receiver front end, the received signals are converted into the intermediate frequency (IF) band. Then they are segmented or channelized into appropriate channel frequency bands by the channelizer and passed into the pre-processor for further pre-processing. Note that only the desired information is passed from the processor to the digital signal processor so that the digital signal processor can be operated at the desired computation rate. It is important to mention that the full electromagnetic bandwidth may be in the range of 2 MHz to 300 GHz, approximately. The bandwidth of the channels may be in the range of 100 MHz to 200 MHz, depending on the application requirements. It should be noted that in addition to greater throughput rates, the channelized receivers can provide substantially simplified architectures in comparison to the architectures available with conventional or anticipated digital signal processing capabilities.

*5.4.2 Limitation on Number of Channel Deployed*

A finite number of channels must be used to meet range and signal-to-noise ratio. However, to meet the stringent range resolution performance, a certain number of channels are needed. Note that the number of channels is limited, if reduction of hardware and signal processing complexity are the principal design requirements. Receiver design engineers feel that to reduce the hardware and signal processing complexity, other technologies must be explored such as the simultaneously intercept concept.

Multi-level architectures with suitable time and frequency multiplexing capabilities provide full frequency bandwidth coverage. Therefore, the probability of intercept of signals is reduced, thereby making the receivers very simple. Note that there is one specific architecture in which the full bandwidth of the electromagnetic environment is divided and subdivided into three levels of frequency segmentation known as bands, sub-bands, and channels. However, in the long run it may be possible to develop very wide bandwidth channelizers with a sub-bandwidth capability of as much as 16 GHz, which is in the frequency from 2 to 18 GHz.

Note that the increased signal throughput rate of the channelized receivers compared to those of other technologies in due to the channelizer capability to simultaneously intercept and/or receive several time-coincident signals. The probability of overlap in time of short pulse width (< than 200 ns, approximately) signals in a given environment for two types of currently deployed receivers (CVR and IFM), along with that of a channelized receiver. This is evident from the data presented in Table 5.1, which shows that the channelized receiver has a fairly large channel bandwidth in the order of 10 to 30 MHz. When the received overlap, incorrect information is reported, which means that the time-coincident signals are present simultaneously within the observed frequency bands of the CVR or IFM receiver band. Note the channelized receiver has a much lower POO, lower than 0.5%, approximately. This is because the channelized receiver has substantially lower bandwidth than the operational bandwidth observed at a given time by a CVR or IFM receiver.

It is critically important to mention that the very high signal reporting rate of channelized spectrum analyzers to the host computer is their most critically limiting performance capability for accurate signal reception in high signal density electromagnetic environments. Note that the pre-processing functions needed to optimize the information rate to the host computer are one of the two most critical requirements which must be addressed to realize the full potential of channelized receivers. Many of the intercepted and channelized date are redundant. A great amount of information can be discarded using appropriate signal pre-processing algorithms so that the host computer can be operated at higher computation rates. In addition, implementation of pre-processing functions and successful implementation of a multichannel monolithic pre-processing circuit for channelizers can be obtained.

It is most desirable to summarize the long-term requirements for both one-channel and two-channel channelizers for the benefit of the readers and receiver design engineers. The long-term requirements include wide bandwidth, high dynamic range, large time-bandwidth product, high frequency resolution, and high DOA resolution. In addition, the physical parameters must be given serious consideration, such as compact and light receiver package and minimum power consumption. Minimum requirements for one-dimensional and two-dimensional channelizers are shown in Table 5.2 [1].

**Table 5.2**  Minimum Requirements for One-Dimensional and Two-Dimensional Channelized Receivers

| ONE-DIMENSIONAL SYSTEM | TWO-DIMENSIONAL SYSTEM |
| --- | --- |
| One-dimensional System | 250,000 receivers |
| Two-dimensional System | 2000 × 2000 channels |

*5.4.2.1  General Requirements*

- Sampling rate: greater than 4,000
- Desirable dynamic range: 80 dB ($10^8$)
- Time-bandwidth product: as high as practical
- Input bandwidth: as wide as possible without spurious response in the band
- Signal processing chips: monolithic technology
- Crosstalk: none
- Sensitivity: as high as practical
- Temporal and spatial noise: as low as possible
- Small channel bandwidth: 1 kHz to 100 MHz
- Small signal response time: 1 N. second
- Receiver size: 82 cm$^3$ (5 in$^3$)
- Receiver weight: as low as possible
- Input power requirement: low power consumption

Due to demanding design requirements for weight, size, and power consumption, the deployment of the latest material and MMIC technologies should be given serious consideration. Furthermore, because of stringent physical requirements of weight and size, the deployment of cryogenic technology can be ruled out. The deployment of the latest packaging technology and advanced or nano-materials requires serious consideration. Performance specification requirements as listed above for the channelizer receivers will be very difficult to meet, unless some relaxation in the packaging size is granted.

For example, to satisfy the packaging size requirement of 5 in$^3$, the receiver package dimensions shall not exceed 1.71 in × 1.71 in × 1.71 in. Figure 5.1 shows the number of components required for the front-end receiver of a non-coherent moving target indicator (MTI) radar. In this case, the minimum number of RF components required is six, namely the mixer, LO, IF amplifier, RF amplifier,

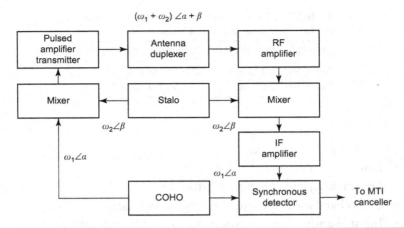

**Figure 5.1** A block diagram of the receiver for a noncoherent airborne moving target indicator (AMTI) radar.

video amplifier, and amplitude detector. Most likely, the front end of a non-coherent receiver can meet the weight and size requirement, provided that strip-line technology and miniaturized components can be deployed. The power consumption, weight, and size requirements for this receiver front barely meet the specification requirements for weight, size, and power consumption.

Figure 5.1 also shows the block diagram of the receiver for a non-coherent airborne moving target indicator (AMTI) radar.

Furthermore, Figure 5.1 shows a front-end receiver block diagram for a pulse-doppler radar or non-coherent radar, which contains seven components such as two mixers, a STALO, a coherent oscillator (COHO), an IF amplifier, an RF amplifier, and a synchronous detector. Note that the output of the detector goes to the MTI canceller. Based on the components required with strict performance requirements, the author feels that it would be impossible to satisfy the stated weight, size, and power consumption requirements. Note that even after combining two or three RF components on a single chip, it is not possible to meet the weight, size, and power consumption requirements for the channelizer receiver, particularly in case of the front-end receiver for the pseudo-coherent radar.

Shown in Figure 5.2 is the block for a receiver of a coherent airborne radar, which consists of ten components, including a STALO, three mixers, an IF amplifier, a COHO, a phase detector, a filter, and a doppler frequency oscillator. In the opinion of the author, it will be

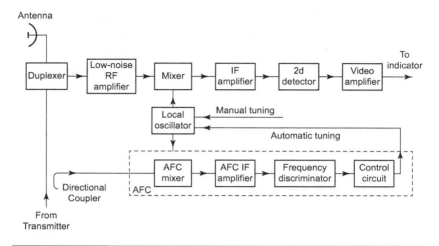

**Figure 5.2** A block diagram for a receiver of a coherent airborne radar, which consists of ten components, including a STALO, three mixers, an IF amplifier, a COHO, a phase detector, a filter, and a doppler frequency oscillator.

difficult, if not impossible, to meet the power consumption, weight, and size specifications requirements.

The critical issues for the channelizers can be summarized briefly as follows:

- Please note that Figures 5.1 and 5.2 summarized the front-end receiver components for one-dimensional channelizers. Obviously, the component requirements and weight, size, and power consumption will be relatively higher.
- As stated before, two or more RF components can be fabricated on a single semiconductor chip, thereby saving weight and size of the combined package weight and size of the combined package.
- Note that if two or three components are fabricated on a single chip, there could be RF leakage and impedance mismatch problems which will degrade the overall RF performance.
- Note that the single-dimensional and two-dimensional performance requirements should be treated as preliminary performance requirements.
- Key signal processing hardware requirements are not included in the front-end receiver components.
- High computational rates are required for the detection of RF signals present in dense electromagnetic environments, and to

process these signals the deployment of one-dimensional and two-dimensional channelizer receivers is a matter of necessity. In order to meet the high computational rate requirement, the use of Fourier transform becomes necessary. This particular approach is considered most suitable for wide-band signal processing.

- The computational powers of analog signal processing and of digital signal processing for channelized receivers need to be defined appropriately.

### 5.5 Computational Power of Analog and Digital Signal Processing for Channelized Receivers

According to the writers of paper [2], the high computational rates required for modern electromagnetic-environmental-signal RF receivers are the main reason for the design and development of channelized RF receivers. To meet the high computational rates, Fourier transform technology is necessary for wide-band processing signal functions. It will be interesting to compare the computation rates for the analog signal processing and digital signal processing techniques. It should be stated that the computational power of analog channelization must be quantified in terms of computer performance capability.

Preliminary studies performed by the author on the computational power of analog signal processing and digital signal processing techniques reveal that the current computational power of analog channelizer technology is greater than that of the current digital technology by an order of magnitude, approximately.

Signal processing experts believe that there is a potential for further improvement and this particular improvement is strictly dependent on channelizer receiver technology due to its inherent parallel signal processing architecture.

### 5.6 Potential Advantages of Parallel Signal Processing Technology

Signal processing engineers believe that the major advantage of this technology is strictly dependent on the interface between the analog channelizer technology domain and the digital electronic computational domain. According to the signal processing experts, the computational power of analog channelization can be considered a major improvement in terms of computer performance. They further believe that both the input and output characteristics or critical parameters must be addressed to evaluate its computational capability.

It is important to point out that the principal channelizer parameters include the input RF bandwidth (B), a time-bandwidth product, and the throughput-time variation. It should be further mentioned that the dynamic range is generally considered to be the electromagnetic signal power range over which the incoming RF signals can be intercepted accurately with a specified low false alarm rate. Note that the effective computational power of a channelizer is strictly dependent on the major channelizer parameters or characteristics. It should be further mentioned that the principal parameters of the channelizer include the input RF bandwidth, dynamic range, time-bandwidth product, and a throughput-time variation or uncertainty. It is critically important to mention that microwave frequencies around 1,000 MHz and lower are used for communication equipment, while frequencies ranging from 1 to 300 GHz are predominantly deployed for radar and ECM applications.

Note that channelization is required for signal processing in high-density electromagnetic environments with specific signal characteristics. Channelization and computational requirements and signal characteristics are mentioned in Table 5.3 [1]. It is critically important to point out that channelization is necessary for signal sorting with specific signal characteristics compatible specified channel bandwidth, as shown in Table 5.3. It should be mentioned that channel bandwidths as narrow as 25 kHz are needed for reception and processing of voice communication signals with frequencies in A, B, and C bands specified in Table 5.2. Note that wideband channelizers bandwidths in the order of 10 MHz, approximately, are required for the interception and processing of radar and ECM jamming signals. In the dense signal environments, several thousands of radar signals or ECM signals may be present in the communication and radar frequency bands at any moment. Various modulation techniques are available, such as amplitude modulation (AM), differential phase shift key (DPSK), frequency modulation (FM), phase shift key (PSK) modulation, pulse code modulation (PCM), and single side band (SSB) modulation. It is critically important to mention that highly selective interception of

**Table 5.3** Environmental Signal, One-Dimensional Channelization, and Computational Characteristics

| FUNCTION | SIGNAL CHARACTERISTICS | | | CHANNELIZER RECEIVER CHARACTERISTICS | | |
|---|---|---|---|---|---|---|
| | FREQ. RANGE (MHZ) | BANDWIDTH (KHZ) | SIGNAL RANGE (DB) | CHANNELIZER (MHZ) | CHANNEL (BW) (KHZ) | COMPUT. (GOPS) |
| Morse (PCM) | 232 | 0.1 | 120 | 30 | 0.1 | 0.55 |
| Voice (comms) | 30–400 | 25 | 80 | 370 | 25 | 5.1 |
| Data (comms) | 30–1,000 | 25 | 80 | 970 | 25 | 15 |
| Radar (PCM) | $100-3\times10^5$ | 10,000 | 60 | 100,000 | 10,000 | 1,300 |

**Symbol definitions**: GOPS = giga operations per second; comms = communications; Comput. = computation; (PCM) = pulse code modulation; (AM/FM) = amplitude modulation/frequency modulation for voice (comms) or (AM/SSB) = amplitude modulation/single side band modulation for voice (comms); (FSK/DPSK) = frequency shift key modulation/differential phase shift key modulation for data (comms) for one-dimensional channelizer.

RF signals both in the communication and radar frequencies rages must be accomplished with high probability.

## 5.7 Computational Requirements for One-Dimensional Channelizers

Note that the computational requirements will be easily managed compared to those required for two-channel channelizers. According to channelizer design engineers, a one-dimensional channelizer is required to meet the needs for communications and radar functions. It should be noted the channelization is most suitable for communications and radar systems because it offers the use of fast Fourier transform computational criteria. Note that equivalent computational requirements range from less than 0.5 giga operations per second (GOPS) for line-of-sight voice communications to 1,000 GOPS for radar signal processing.

### 5.7.1 Tradeoff Studies for Potential Technologies

The author reviewed briefly all possible technologies in dealing with computational methods and the computational requirement with particular emphasis on fast Fourier transform (FFT) computational criteria, channelization coverage, and associated channel bandwidth. Potential channelization receiver technologies will be reviewed, namely SAW, BAW, MSW, and AO receiver configurations.

### 5.7.2 Accurate Measure of Channelizer Computational Power ($P_c$)

Computational power of a digital Fourier transform in complex operations per second can be written as

$$P_{ft} = B\left[\left(\log_2\right)\left(N\right)\right] \tag{5.1}$$

where
  B  is the channel bandwidth, and
  N  is the time-bandwidth product.

$$N = \left[\left(B\right)\left(t_i\right)\right] \tag{5.2}$$

where $t_i$ is the interaction period or time, which is the time interval needed for the detector to register its response to a RF signal in a specified frequency band.

It should be noted that N is also the number of channels in a channelizer.

A more accurate measure of channelizer computational power ($P_c$) in terms of complex operations per second can be given as

$$P_c = \left[ N\left(\log_2 N\right)\right]/ t_T \qquad (5.3)$$

where

   $t_T$   is the channelizer throughput time

It is important to remember that in a matched system, the interaction time ($t_i$) is equal to the throughput time. It should be pointed out that the intersection time or period is of critical importance.

The computational power of computers deployed in the military is typically less than 0.5 million complex operations per second (MOPS). Furthermore, the computational power of current of very high speed integrated circuit (VHSIC) processors is close to 5 MOPS. Due to the deployment of advanced analog to digital converter (ADC) technology, cryogenically cooled Josephson junctions (JJs) made from Nb and aluminum oxide, and rapid single flux quantum (RSFQ) logic circuits, the future VHSIC processors can achieve GOPS greater than 1,320. It appears that computational requirements for multi-dimensional channelization applications are much higher and may be orders of magnitude greater because of the required number of receiver channels, say N channels. Thus, the channelization processing requirement presents a technical channel that is beyond the current state of the processing technology. In addition, a comprehensive environment channelization solution does not appear to be presented by digital technology alone.

A brief overview of various receiver types and channelization technologies seems to indicate that the most efficient channelization technologies are SAW, BAW, MSW, and AO. It should be stated that each of these channelization technologies can be implemented in small volume with a high dynamic range (>60 dB).

Note that the most deployable and popular receiver technologies include superheterodyne (SH), crystal video receiver (CVR), tuned radiofrequency (TRF), instantaneous frequency measurement (IFM), compressive chirp-Z (CCZ) transform, and channelization. Other receiver components such as inter-digital filters and comb line filters were developed to reduce the weight and size of the receivers. But despite the development of unique receiver components, early receiver architectures failed to meet the dynamic range, time-frequency product, POI, POO, and signal processing requirements.

### 5.7.3 Computational Power of Analog Signal Processing for Channelized Receivers

Signal processing experts feel that the current equivalent computational power of analog channelizer technology is relatively greater than that of the current digital technology by an order of magnitude. The experts further believe that there is a potential for further improvement. It should be mentioned that this advantage of channelizer technology is due to its inherent parallel signal processing architecture.

### 5.7.4 Potential for Massive Parallel Signal Processing

It is critically important to mention that the interface between the analog channelizer technology domain and the digital electronic computational domain must be given serious consideration. Note that the computational power of analog channelization could be quantified in terms of computer performance. Furthermore, both the input and output characteristics or parameters of the channelizer must be given serious consideration for the evaluation of the overall computational capacity. It must be stated that the channelizer parameters, namely time-bandwidth product, dynamic range, input RF bandwidth, throughput-time uncertainty, and signal bandwidth must be addressed thoroughly. Note that the input signal dynamic range is considered to be the electromagnetic signal power range over which the signals can be intercepted accurately with a specified false alarm rate. As stated previously, the effective computational power of a channelizer is strictly dependent on the channelizer's characteristics.

### 5.7.5 *Theoretical Limits for the Channelizer Computational Capacity*

Is must be noted that the channelizer processing capability is bounded by the time-bandwidth product and the channelizer throughput speed. As a matter of fact, the time-bandwidth product determines the maximum available level of computational parallelism. In other words, parallel computations can be executed at a frame rate limited by the throughput time. For example, the environments can be channelized with a temporal resolution of 1 nanosecond by a channelizer with a maximum dispersion of 1 nanosecond.

It should be remembered that with the existing channelizer technology, the performance of the channelizer is restricted by the practical considerations of maximum operating frequency and channel density. For example, the channelizer operations through the lower X-band spectrum, that is through the electromagnetic environment frequency range, are equal or less than 10 GHz, as demonstrated with a channelizer consisting of more than 1,000 channels. Essentially, the potential for more than $10^6$ frames per second is provided by the channelizer technology. In other words, an equivalent computational capability for 10 GOPS is possible with the current technology, where the limitations are the throughput time and resolution. It has been stated previously that the current VHSIC processors will be able to meet computational requirements close to 5 MOPS. Semiconductor scientists predict that the next generation of VHSIC technology will satisfy computation requirements close to 150 MOPS [2]. Note this computational capability will be nearly the same as that of the current channelizer technology. The highlights for the computational capabilities of analog parallel channelizer technology and conventional digital signal processing can be summarized as follows:

- The time-bandwidth product with the current VHSIC technology ranges from 10,000 with throughput time of 0.1 seconds to 100 with throughput time of 0.0001 seconds.
- The time-bandwidth product with the current channelizer technology ranges from 100,000 with throughput time of 0.1 seconds to 40 with throughput time of 1 micro second.
- The time-bandwidth product with the next generation of channelizer technology ranges from 1000,000 ($10^6$) with

throughput time of 0.001 seconds to time-bandwidth product 100 with throughput time of 100 nanoseconds.

- The next generation of digital technology offers products identical to those offered by the current channelizer technology, but the new products have slightly greater time bandwidth for throughput times greater than 0.0001 seconds.

- It should be noted that the equivalent computational power of a channelizer is strictly dependent on the channelizer characteristics, namely the time-bandwidth product and throughput time.

### 5.7.6 Analog to Digital Converters (ADC) Device Architecture Using Rare Earth Materials and Low-Temperature Superconductor Technology

An ADC device is the most powerful tool for wide-band signal processing with minimum weight and size. The entire processor can be built on a semiconductor chip with low power consumption. It offers advantages which can be briefly summarized as follows:

- Analog to digital device technology plays an important role in the computational power of the channelizer receiver. ADC devices are designed and developed by HYPRESS, Inc (NY). ADC architecture deploys several hundreds of JJs made of Nb, aluminum oxide, and rapid single flux quantum (RSJQ). Note that these the JJ devices deploy low-temperature superconductor technology [3] with power consumption lower than 100 microwatts. HYPRESSH

- It is critically important to point out that cryogenic technology [4] offers devices best suited for deployment in signal processing for radars and ECM systems. It is important to mention that HYPRESS, Inc have designed and developed ADC devices using cryogenically cooled RSFQ circuits.

- ADC architecture deploys several hundreds of JJs using rare earth materials Nb and aluminum oxide. Note that the quantization architecture of the ADC offers high speed JJ devices that allow sampling rates in excess of 100 GHz and RSFQ logic circuits that allow signal processing in real time. The power consumption in this particular ADC does not exceed 100 microwatts.

- The innovative architecture was first designed and developed by HYPRESS for a 10-bit ADC device that uses more than 800 JJs and has demonstrated a dynamic range equivalent to that of 14-bit ADC.

- This unique LTSC technology offers high resolution, high data rate capability, high sensitivity, ultra-low noise, and power consumption lower than 5 mW per chip at an operating temperature lower than 4.2 K.

- This particular ADC device has demonstrated decimation for the data rate to an optimum sampling rate, which can increase the dynamic range and reduce the harmonic related spurious levels in the pass band of the channel.

- Comprehensive computer programming undertaken by the author on the ADC devices as function of number of bit revealed that sigma-delta ADC architecture offers the highest dynamic range and wider bandwidth as we deploy the higher number of bit. A Nb-based sigma-delta with 10-bit ADC architecture using LTSC technology demonstrated dynamic range in excess of 78 dB and third-order inter-modulation products better than 68 dB based on computer analysis. This analysis further indicated that the extension of flux quantization feedback theory will further improve the dynamic range and instantaneous bandwidth of the ADC devices using LTSC technology.

- Flux quantization provides a mechanically quantum precise feedback mechanism that is not available by other means.

- Superconducting high-resolution ADC architecture based on LTSC technology appears to be the only way to achieve simultaneously low aperture time of the front-end receiver, high internal clock frequency, improved reliability, high data rates over wide band, and high static accuracy with minimum power consumption.

- In summary, it can be stated that the deployment of LTSC technology at 4.2 K cryogenic temperature in the design of an ADC device will achieve low aperture time, high internal clock frequency, high date rate, optimum accuracy, and lower power consumption.

- The entire receiver architecture including mixer, digital filter, ADC and so on can be built on a single semiconductor chip with minimum weight and size.
- The author has applied the LTSC technology concept to ADC architecture for digital radio frequency memory (DRFM). The superconducting DRFM demonstrated excellent ECM effectiveness against a threat deploying appropriate ECM deception modulation technique.

### 5.7.7 Evaluation of Potential Channelization Technologies

The author now wishes to focus on the most promising channelization technologies. Research studies performed by the author seem to indicate that these technologies include SAW, BAW, MSW, and AO technologies. Channelization based on the dielectric resonator (DR) technology is in the research and development stage. Highlights of each technology will be discussed with particular emphasis on its suitability on meeting performance requirements such as bandwidth, dynamic range, time-bandwidth product, and so on.

*5.7.7.1 SAW Technology* Studies performed by scientists reveal that the SAW channelizer technology in the reception RF signals is limited to around 1,500 MHz because of high insertion losses at higher microwave frequencies. Rapid component development in higher frequency band is essential to realize the acceptance in the channelizer receiver approach. In addition, there are lithography difficulties in the fabrication of SAW electrodes. Because of high insertion loss, SAW filters need RF amplifiers to preserve the required dynamic range, thereby receiver package becomes costly, heavy, and power consuming. Due to the involvement of large numbers of filter banks and amplifiers, the reliability problem becomes complicated, and besides there are numerous interconnections. Because of the multiple problems associated with SAW channelizer technology, this particular approach will not be discussed further, unless the problems associated are resolved soon. Despite some problems, SAW technology is preferred by many channelizer designers because of its moderate cost.

However, this channelizer receiver technology is still in the research and development stage.

*5.7.7.2 BAW Technology*   Research scientists feel that this channelization technology has several disadvantages such as poor dynamic range, undesirable cross-talk, heavy and bulky package, and difficult alignment techniques leading to excessive cost and complexity. Because of the above-mentioned problems, this channelization approach is not worth further consideration.

*5.7.7.3 MSW Technology*   It is important to mention that MSW devices can be designed based on forward volume wave (FVW) direction, backward volume wave (BVW) direction, and surface wave (SW) direction. It should be remembered that the major difference between acoustic wave and MSW is that the MSW material is highly dispersive; its velocity changes by orders of magnitude over a bandwidth of a few hundred MHz. The velocity is determined by factors such as saturation magnetization, geometry, magnitude of the bias field, and RF frequency. Under these circumstances, the design of MSW devices is much more complex than that of acoustic wave devices because of the high dispersion property.

MSW device engineers feel that high dynamic range is the overriding consideration in the selection of materials and configurations for frequency channelization. This requires, first, that the insertion loss be low;, second, that the spurious responses be small, and, third, that the power handling capability be reasonably high. These requirements are briefly discussed as follows:

*5.7.7.3.1 Insertion Loss*   Channelizer receiver design engineers believe that several factors must be considered when designing a channelizer with low insertion loss specification. This particular requirement imposes specific demands including efficient excitation of the acoustic or magnetostatic wave, low propagation losses within the acoustic or magnetostatic wave medium, and low channelization loss when the signal is directed from a common input to its appropriate output channel or channels. It is critically important to mention that in case of SAW, BAW, and MSW devices, the input electromagnetic energy must be converted to acoustic or magnetic energy by an electric-to-acoustic or electric-to-magnetic transducer. The type of transducer and its configuration depend on the type of wave used. In order to achieve efficient transducer operation, all transducers must have at least one dimension that is comparable to the wavelength of the excited wave.

*5.7.7.3.2 Transducer Theory*  It should be stated that the active portion of the transducer is an appropriate piezoelectric material in the form of thin film. The thickness of the piezoelectric material determines the excitation efficiency and varies between one-quarter and one-half of the acoustic wavelength in case of BAW transducer. However, the precise thickness of the piezoelectric material is strictly dependent on the geometry and properties of the electrode and transducer material. The transducer can be modeled by a simple RC equivalent circuit, where the capacitance ($C_O$) is the static capacitance of the transducer and the resistance ($R_{rad}$) is the radiation resistance. Note that the radiation resistance determines the power carried away from the transducer in the form of acoustic or magnetic energy for a given input electrical power. For optimum coupling of the acoustic medium, the transducer reactance should be tuned out and the radiation resistance transformed to a circuit-driving impedance, normally to 50 ohms, using a matching network.

In case of SAW devices, the input RF energy is converted to SAW energy by an inter-digital transducer. Transducer designers claim that for generating SAW energy, inter-digital transducer configuration is considered the most efficient. As stated before, a piezoelectric transducer is best suited to launch SAW, MSW, or BAW energy. However, the metal electrode thickness will be different for launching specific wave energy. Note that transducer design configuration and electrode thickness will be different for launching different wave energy.

*5.7.7.3.3 Transducers for Bulk MSW Devices*  Requirements for transducer materials to launch MSWs are slightly different. Note that bulk MSWs are excited by a narrow strip, where the magnetic moments in a thin film of rare earth material YIG are driven directly by the RF magnetic field surrounding the strip. The thin film of

YIG material is placed in the circuit in a region where the field is maximum. Note that the MSW wavelength is a strong function of frequency, and its variations can be in excess of two orders of magnitude within the pass band of the MSW device. It must be noted that as long as the transducer width is less than or equal to one half of the MSW wavelength, the MSW signal could be excited strongly. In this particular transducer, strips are generally very narrow, in the order of 25 microns. The amplitude of the MSW decreases more rapidly with frequency on the low-frequency side compared to the high-frequency side of the pass band. The radiation resistance can be expressed in closed form, but its calculation is more complicated than that for the acoustic wave excitation. However, for surface wave excitation under the simplifying assumption of a ground plane far removed from the YIG film, the MSW radiation resistance can be computed using a very complicated equation involving several parameters, namely permeability of the free space, angular frequency of the MSW, width of the YIG film, transducer width, the elements of the YIG permeability tensor, YIG film thickness, and the MSW wave number, which is equal to $2\pi$ divided by the MSW wavelength. The transducer physical performance can be summarized as follows:

- The piezoelectric substrate must be highly polished to reduce the transducer loss.
- The transducer thickness is one half of the acoustic wavelength for optimum efficiency.
- Surface-acoustic mismatch must be kept to a minimum if high transducer efficiency over a wide band is the principal requirement.
- Optimum frequency for the transducer is defined as transducer material thickness rather than metal line widths and line spacing.
- Transducer electrode thickness and transducer material must be selected for minimum insertion loss and low mismatch loss.
- When acoustic arrays are used, which is most common in BAW and AO transducers, the transducer electrode configuration is generally quarter acoustic wavelength lines separated by quarter acoustic wavelength spaces.

- In case of submicron geometries, one can expect enormous ohmic loss and most difficult fabrication procedure.
- The inter-digital transducer geometry with quarter wavelength finger widths and spacing with alternate fingers tied together by a common bus bar offers the most efficient transducer with minimum insertion loss.
- An acoustic wave is generated by the piezoelectric effect when an alternating potential (AC) is applied between the two sets of inter-digital fingers.
- The equivalent input impedance of a SAW transducer can be
- represented by a parallel radiation conductance and a static capacitance.
- Transducer impedance matching could be a serious problem in terms of loss, cost, and selection of appropriate material for the terminal with unique characteristics.

### 5.7.8 Evaluation of Two Distinct Monolithic Receiver Configurations

(A). Wide-band monolithic receiver configuration [1]

A generic receiver configuration can be used for the implementation of a monolithic receiver capable of full instantaneous waveguide bandwidth performance over the W-band frequency range (75–110 GHz). In this receiver configuration, an RF signal from a wide band antenna is fed to a four-channel multiplexer. Note that the number of channels could be four or eight depending on the acceptance for cost and complexity. In this case, a four-channel monolithic receiver is selected for discussion. It is important to mention that each channel has its own balanced mixer and low-noise IF amplifier. Note that except for the local oscillator (LO), all these components can be integrated on a single chip. The LOs have been developed into a monolithic chip, thereby the entire receiver system contains only three monolithic multi-functional chips.

This particular receiver system has been designed for maximum sensitivity with single- and two-tone inter-modulation distortion, overall receiver spurious-free dynamic range, and multiple-signal handling capability. Note that amplitude and phase distortion keep distortions to a minimum.

It is important to mention that the four-channel approach showed the total number of spurs to be minimum for all channels. Strong spurs are evident in the IF pass band of 8.5 to 17.7 GHz. The receiver design will permit suppression of even harmonics of both RF and LO. The major performance criteria that justify the selection of four-channel receiver configuration are summarized as follows:

- High sensitivity
- Low single-tone inter-modulation
- Wide bandwidth
- Receiver hardware commonality

(B). Multiband, multichannel digital channelizing RF receiver configuration [2]

The developers of the digital channelizing RF receiver claim that this particular receiver configuration will provide excellent performance at minimum cost and complexity. This receiver configuration will provide excellent performance over wide band with minimum weight and size. This multiband, multi-channel digital RF wideband receiver (Figure 5.3) consists of a wide-band ADC modulator and multiple digital channelizer units to extract different frequency bands of interest within the broad digitized spectrum. It should be noted that the

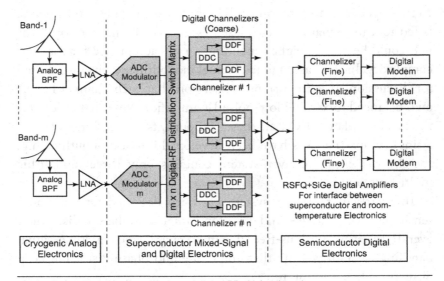

**Figure 5.3**  A multiband, multi-channel digital RF wideband receiver.

single-bit oversampled data, from either a low-pass delta or band-pass delta-sigma modulator, are applied to one or more channelizers, each channel comprising of digital in-phase and quadrature mixers and a pair of digital decimation filters. The channelization is performed in two steps, the first with full ADC sampling clock frequency with RSFQ digital circuits [3], and the second at reduced (decimated) clock frequency with commercial field programmable gate array (FPGA) chip at room temperature [4].

This receiver technology has demonstrated low-pass and band-pass digital receiver technology by integrating an ADC modulator and a channelizer unit on the same chip at clock frequencies up to 20 GHz. Note these 1-cm$^2$ single-chip digital RF receivers contain more than 10,000 JJs [3]. It is important to mention that JJs use low-temperature superconductor technology [4] to meet the stringent performance requirements. Note that the channelizing receiver technology discussed here can be extended to include multiple ADC modulators and multiple channelizer units on a multi-chip module.

## 5.8 Summary

Compressive receiver technology is described as having various major advantages and few disadvantages. This technology is less complex compared to other channelizer technologies. Its slightly inferior performance in terms of dynamic range and time-bandwidth product are marginal. Potential channelized receiver technologies such as MSW, SAW, BAW, and AO have been discussed in great detail. In addition, an overview of technology tradeoff study results is presented for the benefit of the readers and receiver design engineers. Note that each channelizer requires different receiver architecture for specific receiver performance requirements.

Signal processing is the most important and difficult task to derive useful information from the data collected by the receiver in the time allocated. The output signals from the channelizer receiver are in parallel format, which will allow signal processing in a straightforward manner with minimum cost and complexity. The number of channels required for one-dimensional spectral analysis may be as low as 25 or as high as 1,000 depending on the targets and their parameters requirements for classifications and identifications. It is critically important to point out that cost and complexity increase as the number of channels increases.

Principal receiver requirements include signal frequency, amplitude, TOA, DOA, and POI. Some channelizer receivers are required to detect high-density RF signals with high POI, and simultaneous interception of time-coincident signal parameters over wide-signal bandwidth. Note that these wide-band RF receivers are best suited for high-frequency radars, mm-wave communications, ECMs, and ECCMs. Speed of spectrum analysis, signal throughput rate, and accuracy of data collection are of critical importance. It should be noted that simultaneous interception of threat signals and their rapid analysis must be accomplished in the shortest possible time. The very high signal reporting rate of spectrum analyzers to the host computer is their limiting capability for accurate signal parameters in high-density electromagnetic environments. Note that the pre-processing functions needed to optimize information needed for the host computer are one of the two most critical requirements, which must be addressed to realize the full potential of the channelized receivers. Some of the abundant information can be discarded using appropriate signal pre-processing algorithms so that the host computer can be operated at a higher speed.

Component requirements for front-end receivers for coherent pulse radars, non-coherent pulse radars, and pseudo-coherent radars have been described briefly with emphasis on noise figure, dynamic range, and false alarm rate. Frequency stability requirements for STALO and COHO sources are specified. Preliminary studies performed by the author on the computational power capability for analog signal processing and digital signal processing seem to indicate that the computational capability of analog channelizer technology is greater than that of digital signal processing technology approximately by an order of magnitude. Signal processing experts believe that there is a potential for further improvement, due to channelizer receiver technology and also due to its inherent parallel signal processing architecture. It is critically important to mention that the RF input bandwidth, time-bandwidth product, and throughput-time variation are the most critical channelizer performance parameters. Major emphasis has been placed on parallel digital signal processing technology.

According to digital signal processing engineers, ADC device technology can play a critical role in parallel signal processing domain. The ADSC architecture deploys several hundreds of JJ devices made from Nb and aluminum oxide leading to the formation of RSFQ micro-devices. Note that these JJ devices deploy LTSC technology with power consumption lower than 100 microwatts. These ADC devices allow sampling rates in excess of 100 GHz, while the RSFQ logic circuits allow the digital signal processing in real time with power greater than 90 microwatts. ADC device technology operating at LTSC temperatures (around 4.2 K) appears to be the only way to achieve simultaneously low aperture time of the front-end receiver, high internal clock frequency, improved reliability, high data rates over wide band, and improved static accuracy with minimum power consumption.

Various modulation techniques such as amplitude modulation (AM), differential phased shift key (DPSM) modulation, phase shift key (PSK) modulation, pulse code modulation (PCM), and single side band (SSB) modulation are briefly described with particular emphasis on when and where to deploy. Channelization technologies with specific device technologies such as SAW, BAW, and AO. It should be noted that these channelization technologies can be implemented with small size and weight. Furthermore, these devices offer dynamic range in excess of 60 dB. Receiver design engineers can have a choice from the existing receiver technologies such as SH, CVR, TRF, CCZ, and IFM. Other receiver components such as inter digital filters, comb line filters and decimation filters can be deployed if weight, size, and reliability specifications need to be satisfied.

Note that in case of receivers which deploy SAW, BAW, MSW, and AO devices, the receiver engineer faces insertion loss, mismatch loss, and transducer design problems such as critical design of electrodes. The author has evaluated several receiver design concepts, but finally selected a distinct channelizer receiver configuration, namely the multiband, multichannel digital RF channelizing receiver configuration as shown in Figure 5.3. Note that this receiver channelization configuration approach offers high sensitivity, wide input bandwidth, large dynamic range, and large time-bandwidth product. Furthermore, both these receiver channelizer architectures have been discussed in great detail identifying critical design concepts.

In summary, the receiver channelizer deploys a monolithic receiver design concept, ADC devices using LTSC technology, putting two or three components on a single chip, while the entire receiver channelizer including the digital signal processing is located on four chips. Note that both receiver channelizers would meet the weight, size, and power consumption requirements.

# References

1. Martin I. Herman et al., "Multifunction W-band MMIC receiver technology", *Proceedings of IEEE*, Volume 79, No. 1, March 1991, pp. 343–352.
2. Deepnaryan Gupta et al., "Digital channelizing radio frequency receiver", *IEEE Transactions on Applied Superconductivity*, Volume 17, No. 2, June 2007, pp. 430–437.
3. A.R. Jha, *"Superconductor Technology: Applications to Microwave Devices, Electro-Optics, Electrical Machines, and Propulsion Systems*, John Wiley and Sons, Inc, New York, 1998, pp. 176–177.
4. A.R. Jha, *Cryogenic Technology and Applications*, Elsevier Inc, 30 Corporate Drive, suite 400, Burlington (MA) 01803, 2006, pp. 103–104.

# 6

# USE OF RARE EARTH MATERIALS IN MM-WAVE MICROWAVE SYSTEMS AND SENSORS

## 6.0 Introduction

This chapter focuses on the deployment of rare earth materials in the design and development of mm-systems and sensors. The author identifies the rare earth elements, oxides, alloys, and compounds best suited for the development of mm-wave devices, components, and sensors. Significant improvements in performance and reduction in size and power consumption will be given serious consideration. The mm-wave devices and sensors developed using rare earth materials will be most appropriate especially for defense and aerospace applications. Reliability and safety considerations will be given considerable attention.

## 6.1 Identification of mm-Wave Critical Systems, Components, and Sensors

The author will identify the mm-wave devices, sensors, and components which demonstrate potential usefulness in defense, medical, and aerospace applications. It should be stated that the deployment of mm-wave technology will contribute to considerable reduction in weight, size, and power consideration. Now the author will identify further improvements in reliability, electrical and thermal performance, and packaging due to the integration of rare earth materials in the mm-wave devices, components, and sensors. The author will identify the rare earth elements, oxides, alloys, and compounds for integration of mm-wave devices and sensors.

It is important to mention that the reduction in RF component weight and size is strictly to the wavelength of the frequency. In most cases, one will not see an appreciable reduction in weight and size in mm-wave devices due to the introduction of rare earth material technology. Any reduction in weight and size due to deployment of rare earth materials, if any, will be identified.

### 6.2 Typical Rare Earth Elements Widely Used in Microwave and mm-Wave Devices

GaAs, GaN, Si, Co, Sm, Nd, HEMT, and In rare earth elements are widely used in the design and development of RF and mm-wave active devices such as transistors, diodes, detectors, and sensing elements. Note that high-power RF and mm-wave transistors are widely used in the development of high-power RF and mm-wave sources. Indium phosphide diodes are particularly suited for the design and development of mm-wave sources and gun diode oscillators operating at 95 GHz or higher mm-wave frequencies. Later on, specific performance parameters of these unique RF and mm-wave devices will be discussed in great detail. The potential characteristics and unique inherent properties of rare earth materials are summarized in Table 6.1 with particular emphasis on applications in RF and mm-wave frequencies.

There are other rare earth materials which are characterized as hard ceramics; they are widely used as substrates in microwave and mm-wave devices. Some specialized materials have multiple applications in various commercial, industrial, medical, and scientific research applications.

**Table 6.1**    Comparison of Rare Earth Materials Best Suited for mm-Wave Applications

| MATERIAL | MELTING POINT (F) | TENSILE STRENGTH (PSI) | YIELD STRENGTH (PSI) | YOUNG MODULUS (PSI) | THERMAL COND. (W/ cm°C) |
|---|---|---|---|---|---|
| Aluminum | 1,215 | $60 \times 10^3$ | $50 \times 10^3$ | $11 \times 10^6$ | 2.10 |
| Magnesium | 1,200 | $54 \times 10^3$ | $31 \times 10^3$ | $6.5 \times 10^6$ | 1.45 |
| Nickel | 2,610 | $190 \times 10^3$ | $170 \times 10^3$ | $26 \times 10^6$ | 0.65 |
| Titanium | 3,040 | $220 \times 10^3$ | $220 \times 10^3$ | $19 \times 10^6$ | 0.17 |
| Zirconium | 3,355 | $108 \times 10^3$ | $98 \times 10^3$ | $14 \times 10^6$ | 0.16 |

## 6.3 Rare Earth Materials Widely Deployed in Commercial, Industrial, Medical, and Defense Applications

Some distinct rare earth materials which are best suited for multiple applications are shown in Table 6.2 with their applications. The most widely used rare earth materials include Ce, Nd, La, Sm, Y, and Yb. Their important characteristics will be described briefly.

## 6.4 Summary of Properties for Critical Rare Earth Elements

*Ce*

This material is available as a metal, oxide, or compound. Ce is strongly acid, moderately toxic, and a strong oxidizer. Its most commercial applications include metallurgy, glass polishing, ceramics, and catalysts. It is considered the most efficient glass-polishing agent. Ce is widely used in the manufacturing of medical glassware and aerospace windows because of its excellent mechanical strength properties. It is particularly suited for high-quality ceramics for commercial and industrial applications, dental compositions, and phase stabilizers in Zr-based products. The alloys of this material are best suited for automotive power train components. Its nano-particles and

**Table 6.2**   Important Rare Earth Materials and Their Major Applications

| RARE EARTH MATERIAL | APPLICATIONS |
| --- | --- |
| Cerium | Self-cleaning ovens |
| Dysprosium | Fuel control rods in nuclear reactors |
| Erbium | Fiber optic amplifiers, metallurgical processes |
| Gadolinium | Microwave filters, thermal neutron blankets |
| Lanthanum | Microwave substrates, superconductor devices |
| Lithium | Batteries for cell phones, computers, electronic devices |
| Neodymium | Magnet, lasers, colored glasses |
| Palladium | Catalyst, hydrogenation |
| Praseodymium | Lasers, colored glasses for welding operations |
| Samarium | Permanent magnets for motors, generators, TWTAs |
| Thallium | Lasers, photo cells |
| Thulium | Infrared lasers |
| Yttrium | Laser crystals, solid state lasers |
| Ytterbium | X-ray sources |

nanopowders provide ultra-high quality surface area for manufacturing high-performance components. Its oxides are widely deployed in optical coatings and thin-film applications. Its stable and non-radioactive naturally occurring isotopes have medical applications. Its room-temperature thermal conductivity is 0.114 W/cm°C. It has excellent catalytic properties. It is widely deployed in the manufacturing of fiber optic transmission lines, EO devices, and infrared lasers, where cost is the most demanding factor.

*Nd*

This rare earth material has a room-temperature thermal conductivity close to 16.5 W/meter-K and its electric conductivity is roughly $64 \times 10^6$ ohms-cm. The Nd:YAG laser has been deployed for space communication. It has several laser applications including the powerful laser deployment for glaucoma diagnosis. This laser has also been used for other surgical procedures. The long-pulse Nd:YAG laser is best suited for cutting metal plates and rods. Nd is the prime element in the development of multilayer high-power capacitors for power electronics applications.

*Sm*

This particular rare earth element is widely used in the development of permanent magnets for electric motors and generators, which are the critical components of hybrid electric and all-electric automobiles. SmCo magnets are widely deployed as PPMs in the design of high-power TWTAs. These PPM magnets have demonstrated reliable and stable TWTA performance at operating temperatures close to 300°C in the after-burner duration of the fighter-bomber. TWT amplifier designers claim that the deployment of these PPM magnets has significantly reduced the weight and size of the magnetic package. TWT design engineers believe that the PPM magnet configuration offers consistent uniform RF beam over the entire length of the TWTA helix structure. Note that a uniform RF beam is a critical requirement, if high efficiency, maximum gain, optimum output power, and improved reliability are the main requirements.

*Y*

This element is found in abundance in the earth surface containing other rare earth minerals. It is interesting to mention that NASA **has** used this rare earth element for lunar rock sample collection during the Apollo II mission. Y has a silvery metallic luster and is fairly stable in air. Its room-temperature thermal conductivity is around 11.5 W/meter-K and its melting point is roughly 1523°C.

Y oxide is widely deployed for various commercial and industrial applications. It is used to manufacture YIG, which is best suited for microwave and low-frequency mm-wave filters in applications where a sharp cutoff performance in the stop band region is the main requirement. YIG is considered a hard-ceramic material, which is best suited for ultrasonic transmitters and transducers. In case of such transducers, the electrode materials and dimensions of the electrodes are of critical importance. Transducer designers face problems particularly in impedance matching from electrode to piezoelectric material. Note that optimum impedance matching is essential, if low input VSWR and minimum transducer insertion loss are the main requirements.

*Yb*

Yb is best suited for thin-film depositions, sputtering targets, X-ray diffraction, and surface analysis. This rare earth element is widely deployed in several commercial, industrial, and scientific applications. Note that its nano-particles offer high-quality surface areas. It is best suited for optical coatings and thin-film applications. Yb-based lasers offer excellent beam quality, low noise level, and high efficiency particularly at 976 nm wavelength. Its room temperature thermal conductivity is around 34.9 W/meter-K. Its only isotope is Yb-168, which has unique properties and is best suited for many commercial and industrial applications. Its nanoparticles and nanopowders provide high-quality surface areas that are widely deployed in nanotechnology research and in the development of delicate electronic components with unique properties. These components have many commercial, industrial, and scientific applications.

This rare earth element and its oxides are best suited for thin-film deposition techniques using sputtering targets and evaporation materials, optical materials, and other high-technology applications. This rare earth material has a stable and non-radioactive isotope, which is best suited for medical research activities in conjunction with infrared laser sources. Its isotope Yb-168 can be used to acquire physical and chemical analytical techniques and properties such as X-ray diffraction, surface area analysis, and other critical parameters are best suited for industrial applications. Yb is fairly toxic and, thus, it must be handled carefully during rail, sea, or road transportation. Yb-doped fiber optic amplifiers yield optimum performance over wide spectral regions when operated in conjunction with an Yb-based fiber laser emitting at 976 nm wavelength. Note that the fabrication and characterization of Yb-doped Zr-germano-aluminosilicate separated nanoparticle-based fibers yield outstanding optical performance. Laser scientists believe that Yb-based lasers offer excellent beam quality and very low noise levels. Note that fiber optic amplifiers yield excellent performance over wide spectral regions when operated with Yb-based fiber lasers operating at 967 nm wavelength. Comprehensive research and development activities have been focused on finger-mark detection capability, synthesis, photo-acoustic microscopes, and energy transfer techniques. It should be stated that current research and development activities focus on fingerprint detection using Yb O V (4):Er-Yb luminescent upconverting particles; synthesis structure, reactivity of supra-molecular yb-aqua; energy transfer, and enhanced 1.54-nm emission in Er-Yb thin films.

Its insoluble oxides are chlorides that are widely used in metallurgy and chemical and physical vapor deposition, which is ideal for high-performance optical coatings. Other oxides including chlorides, nitrates, and acetates are available in soluble form. It is critically important to stress that soluble oxides are best suited for the development of chemical compounds for industrial applications.

*Gd*

Rare earth material scientists claim that Gd's atomic structure, ionization energy, room-temperature electrical conductivity, and thermal conductivity are best suited for MRI and other medical diagnosis applications. This rare earth element is particularly useful as an effective injectable contract agent for patients undergoing MRI procedures.

It is extremely important to mention that its high magnetic moment can significantly reduce relaxation times, leading to improved signal intensity. Gd can be used as a host for X-ray cassettes and in scintillator materials for computer tomography and other medical procedures.

Gd is available as an oxide or a compound. This rare earth element comes in rods, wires, pallets, and granules. Its nano-particles and nanopowders offer excellent surface polishing agents, and after polishing the surface the object can be examined with a microscope under high-intensity light for any crack or chip in the surface. Its oxides are considered excellent for high-quality optical coatings and thin films for specific applications. Note that thin films are best suited for microwave and mm-wave superconducting devices when operating at cryogenic temperatures. These thin films of suitable rare earth materials are widely used for superconducting microwave filters, where minimum insertion loss, high-quality factors, reliability, and consistent device performance are the principal performance requirements.

Compounds of this element can be manufactured as per specifications. Note that Gd compounds have no specific biological role. This element is very toxic and, therefore, it should be handled with extreme care during transportation. However, its isotopes are stable and non-radioactive. Its isotopes are best suited for medical research studies and analytical investigations such as X-ray and precision surface area examinations. Its room-temperature electrical conductivity is around 141 megohms-cm and its thermal conductivity is close to 10.5 W/meter-K. Research and development activities on this element have focused on medical treatments and diagnosis procedures. Medical research scientists have focused their research and development efforts on medical treatments and diagnosis of various diseases. Current research and development activities are mainly focused on the following medical areas [1]:

- Feasibility studies using MR enterography for terminal inflammatory activity, especially in children
- MRI findings of the parotid gland
- MRI and CT precision evaluations of congenital pulmonary vein abnormalities
- MRI characterization of progressive cardiac dysfunction

- Ultra-sonography, magnetic resonance imaging, and computer tomography for specific medical diagnosis and treatments
- Effectiveness of combined magnetic resonance imaging and contrast-enhanced computer tomography for specific medical treatments and diagnosis procedures
- Potentiality of gastric motility drugs for prostate cancer imaging
- Precision assessment of distribution and evolution of mechanical dyssynchrony of a myocardial case
- Hyperpolarized spectroscopy to detect tumors in early stages

It is critically important to point out that Gd oxide sputtering targets, Gd-selenide sputtering targets, Mg-Gd sputtering targets, Gd-telluride sputtering targets, and Zr-Gd sputtering targets play key roles in several commercial and industrial applications. Note that ultra-thin Gd thin foils are best suited for scientific research applications, where reliability and precision measurements are the main requirements.

*Er*

Research and development studies performed on diode-pumped solid state lasers indicate that this rare earth element is generally selected to absorb the pump light and Tm is used to transfer the optical energy from Er to Ho as illustrated by Figure 6.1. An investigation on doping case using Ho host crystal seems to indicate that Ho, which emits at 2.05 nm, is not considered an efficient optically pumped laser. However, the use of sensitizer materials can increase the efficiency of the pumping. Note that when a host crystal such as YAG is co-doped with another suitable rare earth material such as Er or Ho, significant efficient laser action has been demonstrated due to energy transfer from one ion to another. Er plays an important in the design of lasers and EO sensors.

Laser scientists believe that appropriate doped rare earth crystals must be deployed in the design and development of IR lasers emitting at the desired wavelengths. Note that when dealing with Nd-doped YAG is used, thermal broadening effects can be observed. Research studies performed by the author reveal that the most popular rare earth

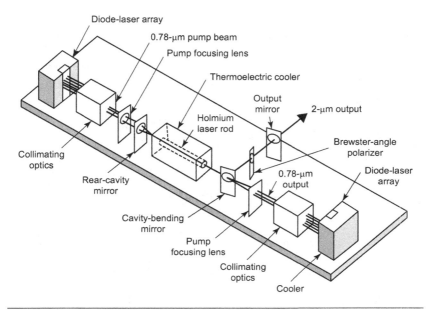

**Figure 6.1**   2.9-micron eye-safe laser configuration for a Ho laser using a solid state pumping scheme.

elements used in the design and development include Nd, Y, Er, Ho, Eu, Tm and YAG. The block diagram of a solid state laser Ho:YLF can be seen in Figure 6.2, complete with water cooler, optical mirrors, collimating optics, solid state pump, Ho:YLF laser rod, cryogenic cooler, AlGaAS diode array, and important accessories. The output wavelength of the laser is 3.95 nm. Note that a Tm:Ho:YAG laser emitting at 2.1 nm is considered the most ideal laser system for eye-safe eye-related treatments. The performance parameters and potential applications of laser are summarized in the following paragraphs.

## Th

Uranium and Th are considered the basic rare earth elements for nuclear reactors to generate electrical energy. Th is available as a metal, oxide, or compound. Th is referred to as a lanthanide rare earth material and widely deployed for nuclear power plant applications. This material comes in the form of foil, rod, or sputtering target. Th compounds come in the form of sub-micron and nanopowder. This material is widely used as a tungsten coating in electronic components

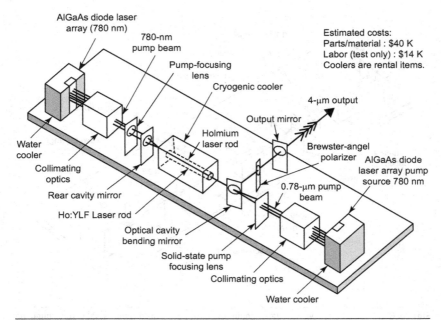

**Figure 6.2** A block diagram of a solid state laser Ho:YLF.

due to its high emission characteristics. Its oxides are widely deployed in advanced EO devices, because of its high refractive index value. It is also used in several other high-temperature glass applications such as in the mantles of lamps and to produce crystal growth crucibles.

The molecular formula, safety data, research information, and properties are readily available in university science libraries. This element is available in form of rod, pellet, wire, and granule. The oxides of this material are available in form of powders and dense pellets and are best suited for optical coatings and thin-film applications. Devices with Th coatings are widely used applications, where high-quality performance, reliability, and device longevity are the principal design requirements. Its insoluble oxides are fluorides which have potential applications in metallurgy, optical coatings, chemical processes, and physical vapor deposition (PVD) techniques.

Research and development activities undertaken on Th on an international level can be briefly summarized as follows:

- Synthesis and characterization of Th sulfates
- Synthesis, structural activity, and computational studies on Th oxides

- The scientific discoveries of uranium-235 and symmetric fission
- Matrix infrared spectroscopic and density functional theoretical investigations on Th and uranium atom reactions with dimethyl ether
- Interaction of Th with nitrate in aqueous solution
- Use of liquid Th for nuclear power reactors
- A cryogenic beam of refractory, chemically reactive molecules with expansion cooling
- Gamma-ray laser capable of emitting in the optical spectrum range
- Gamma-spectrometric analysis of high-salinity fluids
- Screening of plant species for phytoremediation of uranium, Th, Ba, Ni, Sr, and lead-contaminated soils from a uranium mining tailings depository in South Africa
- Background radiation and individual dosimetry in the coastal region of the Tamil Nadu state of India
- Dust concentration analysis in non-coal mining areas and its impact on human health
- Exposure evaluation based on measurements undertaken for occupational hygiene at research laboratories in Poland from 2001 to 2005 [2]

## 6.5 Applications of Rare Earth Oxides $ZrO_2$ and $Y_2O_2$ in High-Power Fuel Cells

Fuel cell scientists and engineers were hesitant to pursue extensive research and development activities on high-temperature, high-power fuel cells due to a lack of appropriate rare earth materials and technological difficulties. In the early 2000s, American Elements fuel cell scientists concluded that the deployment of specific rare earth materials for cathodes, anodes, and electrolytes may be required to produce a highly compatible thin-film electrochemistry with outstanding thermal expansion-matching capability. They felt that thermal expansion-matching capability would yield optimum fuel cell performance in terms of power output, efficiency, safety, and reliability.

With thermal expansion-matching capability, research on high-power fuel cells started around 1960s. Fuel cell research scientists believe that high-power fuel cells require operating temperatures close to 1,000°C or higher. Note that at such temperatures, conventional electrolyte materials such as semi-solid molten electrolytes or aqueous electrolytes will not meet the output power and reliability requirements. In early 1960s, several fuel cells were designed, developed, and evaluated. Those early fuel cells included 10 fuel cells using aqueous electrolytes, 6 fuel cells using molten electrolytes, and 3 fuel cells using solid electrolytes. Extensive and comprehensive experimental performance evaluations were made in terms of output power, electrochemical efficiency, reliability, and safety. These tests concluded that fuel cells using solid electrode technology would meet the above performance specifications requirements.

Critical rare earth materials and principal performance requirements for electrolytes, cathodes, and anodes will be briefly summarized. Because of the high-temperature requirements of the materials needed in the design and development of high-power fuel cells, the development of fuel cells was put on the back burner during the 1960s and 1970s. However, some scientists in Europe continued their research and development activities in during that time frame.

It is interesting to mention that fuel cell scientists at General Electric and Westinghouse Electric Corporation were deeply involved in the R and D activities on high-temperature rare earth materials. GE fuel cell scientists used Zr oxide ($ZrO_2$) as a solid electrolyte with no doping. When the scientists doped the Zr oxide with Y oxide ($Y_2O_2$), significant improvement was noticed in fuel cell efficiency from 35% to 48% and in the cell life from 2,100 to 3,600 hours with no compromise in safety and reliability, according to GE scientists. They claim that when using a higher level of doping with Y oxide, further improvement can be expected in the cell efficiency. Note that, so far, the efficiency improvement was observed due to the deployment of Y oxide in the electrolyte. Research and development activities on fuel cell continued using various rare earth materials for cathode and anode elements.

Thereafter, fuel cell scientists focused on the deployment of rare earth oxides in the design of cathode and anode terminals of the cell. They discovered that, due to the presence of temperatures close to 1,000°C or higher, semi-solid molten electrodes or aqueous electrodes will not be able to meet the high-temperature requirements in the fuel cells.

## 6.6  Potential Rare Earth Oxides Best Suited for Electrolytes

The author has reviewed potential rare earth materials which might qualify as electrolytes. Studies undertaken on high-temperature rare earth materials indicate that stabilized Zr oxide (YSZ) can make a robust electrolyte material which is ionically conductive and can operate at a wide range of partial pressure. This electrolyte material can operate at 1,000°C or higher; this has no impact on its performance. Comprehensive material research studies suggest that Ce oxide stabilized with Gd oxide (GDC), Ce oxide stabilized with Y oxide (YDC), and Ce oxide stabilized with Sm oxide (SDC) for a class of electrolytes with higher ionic conductivities and higher operating temperatures exceeding 1,000°C. Note that these electrode materials operate at narrow partial pressures and will electronically conduct if operated at lower partial pressures. SCZ electrolyte material is roughly three times more ionically conductive than YSZ electrolyte material and operates very efficiently at temperatures slightly lower than 1,000°C.

GE fuel cell scientists designed and developed a fuel cell using natural gas as fuel. The gas was enclosed in a heating jacket. The operating temperature was measured as 1,093°C when the fuel call was operating at full capacity. It should be mentioned that the natural gas decomposes carbon and hydrogen at this operating temperature. The classification of fuel cells is based on the type of materials used for the electrolyte. Some fuel cell designers characterize classification based on all three materials, namely electrolyte, cathode, and anode.

The fuel cell classifications based on electrolyte materials can be summarized as follows:

- Alkaline electrolytes
- Phosphoric electrolytes
- Molten-carbonate electrolytes
- Solid-polymer electrolytes
- Solid-oxide electrolytes
- Aqueous electrolytes

Various fuel cell configurations using aqueous or liquid electrolytes or methanol fuel have been investigated in terms of efficiency, safety, cost, and performance. Thermodynamic-based fuel cell voltage (E°) and electrochemical efficiency for various fuel cells are summarized in Table 6.3.

**Table 6.3**   Cell Voltage and Electrochemical Efficiency for Various Cell Types

| CELL FUEL | VALANCE | E° (VOLT) | IDEAL CELL EFFICIENCY (%) |
|-----------|---------|-----------|---------------------------|
| Hydrogen | 2 | 1.229 | 83.4 |
| Carbon oxide | 2 | 1.066 | 90.4 |
| Formic acid | 2 | 1.480 | 98.6 |
| Methanol | 6 | 1.214 | 96.7 |
| Methane | 8 | 1.060 | 91.8 |
| Ammonia | 3 | 1.172 | 88.5 |
| Hydrazine | 4 | 1.558 | 96.8 |

Comprehensive research studies performed on fuel cell structures indicate that the direct methanol fuel cell (DMFC) configuration deploys a proton exchange membrane (PEM) structure. Specific performance capability and important features are described as follows:

- A fuel cell generates electrical energy by an electrochemical conversion technique, which can be replenished with minimum time and effort.
- Latest studies seem to indicate that the fuel cell configuration shown in Figure 6.3 is best suited for portable, medium-power applications because the $H_2O_2$ technology offers the most compact device design, improved reliability, and significant reduction in weight and size.

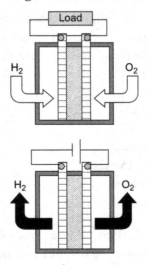

**Figure 6.3**   Fuel cell architecture involving Si-based $H_2O_2$ fuel cell technology operating at room temperature.

- It is critically important mention that methanol fuel is relatively inexpensive and readily available with no restriction on fuel quantity.
- It is interesting to mention that German scientists have designed the fuel cells to operate at ambient temperatures and pressures.
- Note that German scientists also designed and developed double skeleton catalyst (DSK) fuel cells using methanol and catalytically different electrodes, while Swiss scientists designed mono-skeleton catalyst (MSK) fuel cells using cheap fuel (hydro-carbons) and electrochemically active metal electrodes.

## 6.7  Most Common Elements Deployed in Describing Fuel Cells

- Electrode: this term is generally used to identify the positive terminal of the fuel cell.
- Anode: this is the negative of fuel electrode.
- Cathode: this is known as the oxidizing electrode that accepts electrons from the external circuit and oxygen is reduced in the process.
- Membrane electrode assembly: this is a laminated, sandwich-type assembly, which contains two porous electrodes separated by an ion-conducting polymer electrolyte. The catalyst is a part of this assembly.
- PEM: a polymer film is used to block the passage of gases and electrons while allowing the hydrogen ions known as protons to pass.
- Reformer: a small chemical reactor carried on the board some fuel cell vehicles used to extract hydrogen from the alcohol or hydrogen fuel.

## 6.8  Requirements for Cathode

The latest material science studies performed by the author seem to indicate that La-based materials might meet the temperature requirements. The following rare earth materials might meet the thermal requirements for the Pervoskite Cathode Materials [3]:

**Table 6.4**   Thermal and Mechanical Properties of Rare Earth Materials Best Suited for Fuel Cells

| RARE EARTH MATERIAL | MELTING POINT (°C) | YOUNG'S MOD. (G. PSI) | TENSILE STRENGTH (K. PSI) |
|---|---|---|---|
| Chromium | 1,890 | 50–75 | 100–500 |
| Berrylium | 1,278 | 68–73 | 88–95 |
| Cobalt | 1,480 | 65 | 65–75 |
| Molybdenum | 2,500 | 32 | 150 |
| Nickel | 1,750 | 22 | 70 |
| Platinum | 1,785 | 21 | 51 |
| Palladium | 1,550 | 17 | 39 |
| Strontium | 770 | 16 | 32 |
| Titanium | 1,675 | 30 | 210 |
| Tungsten | 3,400 | 59 | 500 |

- La Ca magnetite
- La Sr ferrite
- La Sr cobaltite
- La Sr chromate
- La Sr gallate magnetite

It is interesting to examine the critical thermal properties of some materials. Table 6.4 offers the critical thermal properties of rare earth materials deeply involved in the design of cathodes.

### 6.9  Rare Earth Elements and Crystals Best Suited for Solid State IR Lasers

There are only a few rare earth materials that are suitable for IR lasers. In addition to melting point temperature parameters, semiconductor characteristics must be given serious consideration, such as stable optical laser beam, beam stability, and DC-to-IR conversion efficiency. Since the optical power from a single optical diode is very limited, one has to consider the diode-pump laser technology concept or a lamp-pump laser technology approach to achieve higher IR power output at 2.1-micron or 2.9-micron or 3.9-micron wavelength.

In case of a diode-pumping scheme, a laser input bar consisting of GaAs or InGaAs diodes, operating in parallel configuration in an optical cavity, is used to pump an appropriate rare earth crystal such as Er:YLF or any appropriate rare earth crystal to achieve IR power at 2.0 microns or higher, as shown in Figure 6.1.

*6.9.1 High-Power Coherent Laser Source*

In the case of a high-power quantum well laser, the lamp-pumping scheme will be found most appropriate depending of the IR output power requirements. Note that for lower IR power levels, a diode-pumping scheme should be preferred, if moderate IR power level, high reliability, small package, and high beam stability are the principal design requirements. The DC power supply will drive the InGaAs or AlGaAsP quantum well arrays which will generate IR energy at 780 nm. These diodes can be arranged in parallel in an optical cavity, if higher IR driving energy is desired. A water cooler should be used to keep the cavity temperature around 10°C to maintain higher conversion efficiency. The output of this cavity will go to another optical coupler. The output of this coupler will drive the Tm:Ho:YAG laser rod capable of generating IR energy at 2.9 microns, which will drive the Ho:YLF laser rod IR power close to 1 W at 3.94-micron wavelength as shown in Figure 6.2. Note that both the power supply and the water cooler are equipped with precision control circuits to maintain the IR output power level and cooling temperature at the optimum values, and to maintain beam stability and conversion efficiency within the desired specifications.

The overall dimensions of the laser package, weight, cost of the AlGaAS array, and minimum cost the Tm:Ho:YAG laser rod and Ho:YLF laser rod are shown in Figure 6.1. These estimates for the physical parameters were true as of 2013. It is fair to assume that the cost of laser rods now will be higher by 50% or more.

*6.9.2 Use of Rare Earth–Based Quantum Well Diodes for Optical Lasers*

It is important to point out that InGaAsP/InP or InGaAs/InGaAsP semiconductor laser diodes with quantum well (QW), double quantum well (DQW), or multiple quantum well (MQW) will offer higher optical power levels with improved differential quantum and conversion efficiencies. These diodes deploy rare earth materials such as indium (In), gallium (Ga), and phosphor. It should be stated that with epitaxial material layers grown on P-substrate, excellent lifetime characteristics can be achieved. Inclusion of strained-layer InGaAs quantum well into an InGaAsP heterostructure will allow this semiconductor laser diode to generate IR energy over 1.55 to 2.1-micron range without cryogenic cooling.

It should be noted that these quantum well diodes with InGaAs-DQW active layers can deliver optical power in the order of several hundred mW per facet. Computer analysis performed by the author seems to reveal that the room-temperature total differential quantum efficiency (TDQE) can be as high as 46% and the threshold current as low as 26 mA at room temperature. However, under cryogenic cooling down to −10°C, the threshold current is reduced to 14 mA, approximately. Research studies performed by the author indicate that the FWHM laser beam width in the perpendicular plane remains constant regardless of the optical power level. Theoretical analysis indicates that a small aspect ratio of the ridge waveguide significantly simplified the output optical cavity design, thereby realizing a substantial reduction in the weight, size, and cost of the laser assembly.

It is critically important to point out that the thickness of the heavily strained active regions imbedded within the InGaA P heterostructure determines the emission wavelength, whereas the mole fraction of the In and Ga in the quantum well determines the internal transparent current density (ITCD). Note that the heterostructure laser diode employs the binary, which offers the highest thermal conductivity and the heat transfer efficiency of the diode junction. It should be noted that heat transfer efficiency, drive current, and operating temperature determine the optimum optical power output. Typical TDQE is dependent on the reflectivity and the room-temperature optical power, while DQE is strictly a function of drive current. This laser architecture is for a 2-micron semiconductor laser system. It is interesting to mention that a tetravalent chromium laser crystal such as $Cr^4:Ca_2 GeO_4$ is best suited for a tunable laser operating over 1.3- to 1.6-micron range, which has a potential application in covert optical systems operating in heavily electronically defended environments.

This laser configuration offers improved total differential quantum efficiency, low threshold current, non-cryogenic operation, high output power, improved reliability, and eye-safe operation. Its potential applications include eye-safe surgical tools, laser-based radars (LIDARs), optical communications, IR countermeasure techniques, and directed energy weapon systems. It should be mentioned that the typical differential quantum efficiency for a quantum well laser diode is about 45% at room temperature, which rises above 65% under cryogenic conditions. Note that its room-temperature injection efficiency is about 70%.

## 6.10 Deployment of Rare Earth Materials in mm-Wave Radiometers and Radar Systems

Under this heading, the author wishes to describe the performance capabilities and operational parameters of some mm-wave systems employing rare earth materials. The author will describe the vital capabilities of a mm-wave radiometer sensor, 95 GHz airborne radar, and other mm-wave sensors that have been designed and developed using rare earth elements. Note that attenuation or atmospheric absorption is much higher at mm-wave frequencies.

### 6.10.1 Description of 90 GHz Radiometer and Its Capabilities

This particular sensor is a passive system and is capable of operating at mm-wave frequencies. The author will describe the performance capabilities of a 90 GHz radiometer and its outstanding design features. This particular radiometer was designed and developed by the Georgia Tech Research Institute in Atlanta (GA). The system deploys two radiometric receivers, located at two positions in the direction of travel. This system detects the microwave energy radiated or reflected from the terrain which contains grass, crops, plowed fields, and man-made objects. Note that, as the system moves along, signal levels are detected in each mm-wave receiver. The receivers' outputs will be cross-correlated and the time delay between the two signals will determine the speed. No signals will be transmitted, thereby avoiding problems relating to emission of radio frequencies. The system is carried on a ground vehicle such as a tractor, light tank, or military sub-type vehicle. Note that most of the detectors used in mm-wave radiometers are made from rare earth elements such as In, Cd, telluride, and so on. Note that the down-looking mm-wave directional antenna moves horizontally over the terrain and man-made objects such as buildings, ground-based moving objects, and other fixed targets. The varying signal received by the antenna is related to the emissivity and reflectivity of the ground-based objects. The signals are received by two antennas: one being delayed with respect to the other. Note that the amount of delay would be related to the separation of two antennas and the speed of the moving vehicle on the ground. Since the separation between the antennas is fixed, measuring the delay would permit the operator to determine the speed of the moving vehicle. Recent measurements undertaken by the engineers at Georgia Tech in Georgia indicate that accurate speed measurements of ground-based moving targets can be made with this mm-wave radiometer. It is interesting to mention that deployment of mm-wave radiometers is best suited for airborne surveillance and surveillance systems, where high-resolution images and detection of high value targets are the principal requirements.

### 6.10.2 Performance Capabilities of mm-Wave Transmitters Using Rare Earth Materials

Comprehensive research studies undertaken on mm-wave transmitters indicate that extended interaction oscillators (EIOs), TWTs, and gyrotrons could be used as radar transmitters. But the weight, size, and power requirements would be not be acceptable to the radar designer. The radar engineer would prefer solid state transmitter sources such as Gunn diodes or IMPATT diodes. Note that physical parameters for solid state transmitters, as shown in Table 6.5, would be found more desirable for radars operating at higher mm-waves.

### 6.10.3 Critical Parameters of mm-Wave Radar Transmitters

The mm-wave (MMW) radar transmitters have unique characteristics which are briefly summarized in Table 6.5.

### 6.10.4 Operating Parameters of 95-GHz Airborne Radars

**95-GHz radar parameters** assuming the circular antenna size of 1 meter, 1 kW peak power for the transmitter with a pulse width of 0.1 microseconds, and a PRF of 10,000 pps, and neglecting scattering losses due to rain, are as follows:

**Table 6.5** Critical Parameters of MMW Radar Transmitters

| TRANSMITTER TYPE | TYPICAL OPERATING FREQUENCY (GHZ) | | | TECHNOLOGY RARE EARTH (RE) |
|---|---|---|---|---|
| | 94 | 140 | 220 | |
| Type of power source: | | | | |
| TWT (kW) | 1–2 | 0.15–0.20 | .05–.06 | Uses RE magnet |
| Gyrotron (kW) | 20 | 9 | 1 | Uses RE magnet |
| GUNN or IMPATT DIODE (W) | 12 | 2.6 | 0.5 | Uses RE (InP) |
| ANTENNA SIZE 1 meter (assumed) | | | | |
| 3-dB beam width (°C) | 0.24 | 0.16 | 0.1 | |
| Antenna gain (dB) | 57 | 60 | 64 | |
| One-way attenuation (dB/n. mile) | 0.49 | 0.55 | 1.31 | |
| One-way rain (4 mm/hr) attenuation (assuming one-way (dB/n. mile) and neglecting scattering due to drop size) | 7.5 | 8.5 | 9.5 | |

## Tactical Radar Performance Capabilities

| | |
|---|---|
| Operating modes | Semi-active missile guidance, vital air-to-air and air-to-ground modes |
| One-way antenna gain | 57 dB |
| 3-dB beam width | 0.23°C |
| Transmitter source | TWT |
| Signal processor type | High-speed computer |

**Other possible radar modes of operation beside missile guidance**
Search, detection, target tracking, precision ground mapping, and ground moving targets.

*6.10.5 Critical Tactical Radar Performance Requirements*

Tactical radars must be fully equipped to meet the following modes and precision functions:

- Radar transmitter and receiver design requirements must be suitable for coherent capability
- Pencil beam of the antenna is needed for tracking airborne targets
- Cosec$^2$ antenna beam is necessary for ground mapping irrespective of aircraft altitude

- Specific transmitter pulse width provisions must be provided for tracking airborne targets and for wide-area ground mapping capability
- High-pulse compression ratio is needed for precision target tracking and ground mapping
- Transmitter must satisfy the required jamming power requirements. Note that atmospheric attenuation will reduce the amount of effective radiated power (ERP) of the jammer needed to jam the enemy radar in a self-protection ECM. However, in case of stand-off ECM missions, atmospheric attenuation may magnify the ERP required from a jammer.
- To meet the ground clutter requirements during the mapping mode, the 3-dB antenna beam must be lower than 0.25 degrees and the transmitter pulse as narrow as possible.

### 6.11 mm-Wave Radiometers and Their Applications

MM-wave radiometers are best suited for surveillance of tactical targets and to obtaining moderate-resolution images of ground-based objects or targets. This mm-wave sensor can be deployed for a variety of remote sensing applications and produces high-quality images through clouds, smoke, and dust, when other IR sensors are not usable. This system can be installed on a small aircraft or helicopter to generate near real-time, moderate-resolution images of ground-based structures or targets of great interest. In addition, it can also be used for variety of remote-sensing applications, such as measurement of snow cover, surface moisture, vegetation type, extant mineral types, surface moisture, and thermal inertia. This sensor can be deployed to map fires in remote controlled areas which are difficult to access. It can observe different physical phenomena, which makes it a valuable visual and IR imaging sensor.

When a mm-wave radiometer observes a ground scene from above ground, the received signal temperature is composed of emission from the objects present in the antenna beam reflected sky emission, and the emission from the atmosphere below the radiometer. The received temperature can be the sum of various temperature sources. The resultant received temperature can be expressed as,

$$T_{received} = \left[ T_{object} + T_{background} + T_{reflected\,from\,sky} + T_{atmosphere} \right] \quad (6.1)$$

Note that all temperatures are expressed kelvin. It has been assumed that the power received by the radiometer is within the main antenna beam, which is defined by the half-power antenna beam that is typically assumed as being 0.5 degrees or less. Note that this is a good assumption for radiometer antennas with beam efficiencies greater than 90 degrees.

It should be noted that the received temperature from an object or target in the antenna beam is equal to the product of object temperature and its emissivity, multiplied by the ratio of the object area to the antenna beam area and reduced by the atmospheric attenuation.

The object temperature can be expressed as,

$$T_{object} = [\varepsilon T_\eta / L_a] \quad (6.2)$$

where $\varepsilon$ is the emissivity of the object, $L_a$ is the atmospheric absorption below the radiometer, and the parameter $\eta$ is the ratio of the object area to the main beam area. This parameter can be written as,

$$\eta = [A] / [\pi/4] \left[ R \tan \vartheta_{beam} \right] \quad (6.3)$$

where,

A   stands for object area (meter$^2$),
R   stands for range to object (meter), and
$\vartheta_b$   indicates the half-power beam width.

The reflected sky emission temperature is equal to the product of the radiometric sky temperature and the resultant emissivity multiplied by the ratio of the product area to the antenna beam area and reduced by the atmospheric absorption ($L_a$). Thus the reflected emission temperature can be written as,

$$T_{reflected\,sky} = \left[ (1-\varepsilon) T_{[sky\,\eta]} / L_a \right] \quad (6.4)$$

where $T_{sky}$ is the radiometric sky temperature in kelvin. This radiometric sky temperature can be expressed approximately as,

$$T_{sky} = T_{atm} \left[ 1 - 1/L_z \right] + \left[ T_{ch} / L \right] \quad (6.5)$$

where $T_{atm}$ is the mean atmospheric temperature, $L_z$ is the total zenith absorption loss, $T_{ch}$ is the cosmic background temperature, and L is the channel temperature loss in Kelvin. The received signal temperature from the background surrounding the object or target can be expressed as,

$$T_{background} = \left[ \varepsilon_g T_g + (1 - \varepsilon_g) T_{sky} / L_a \right] \cdot \left[ (1 - \eta) \right] \qquad (6.6)$$

where,
$\varepsilon_g$    is the emissivity of the ground and
$T_g$    is the physical temperature of the background.

The atmospheric temperature below the radiometer can be written as,

$$T_{atmos} = T_a \left( 1 - 1 / L_a \right) \qquad (6.7)$$

where,
$T_a$    is the physical temperature of the atmosphere below the radiometer.

Besides these equations, one must realize the prevailing atmospheric conditions in dealing with the overall performance of the radiometer.

### 6.11.1 System Description and Operating Requirements of the Radiometer

The radiometer description and operating requirements can be summarized as follows:

- All temperatures are expressed in kelvin.
- Under various atmospheric conditions for reflective objects one must assume low emissivity ($\varepsilon = 0.1$) and for absorbing objects one must assume high emissivity ($\varepsilon = 0.8$).
- When the object fills the antenna beam ($\eta = 1$) and the signal is received from a range of 750 meters, approximately.
- It should be remembered that reflective targets or objects such as metals or water surface appear colder than the surrounding high-emissivity ground such as soil or vegetation. Under these conditions, the reflective objects stand out clearly from

the surrounding background with a temperature difference or signal-to-noise ratio.

- In case of a small mm-wave sensor for the detection of reflective objects in cloudy weather, it was determined that the 98-GHz window provided the best signal-to-noise ratio at any operating frequency available. However, at the other frequency windows such as 140 and 220 GHz, the gain of the smaller antenna beam width was cancelled by the larger atmospheric attenuation in the cloudy weather.

*6.11.2 Radiometer Scanning Capability*

- As far as the radiometer antenna scan capability is concerned, a lightweight flat reflector antenna can be scanned mechanically across-track over +/− 20 degrees at a 4-hertz scanning rate.
- Studies performed by the author on the scanning requirements indicate that a compact two-axis scanner, controlled by a micro-computer with inputs from rate and angle gyros, can be used to scan the antenna and to correct for aircraft movement due to turbulence. This provides line-of-sight stabilization without the use of a conventional stabilized platform, which would make the system very heavy, bulky, and costly.
- A pointing stability of 0.1 degrees (17.6 mrad) eliminates pixel smear and image distortion.
- A signal from the ground is reflected from a flat scanning mirror onto a flat offset lightweight antenna, which focuses the received signal on to the radiometer feed horn.
- Note that the offset parabolic must be designed for high beam efficiency greater than 90% and for ultra-low side lobes greater than 35 dB.
- Two linearly polarized 98-GHz superheterodyne radiometers will be required, if polarization information is needed.
- These two radiometers deploy balanced mixers using beam-lead diodes to achieve better radiometer performance.
- A Gunn diode local oscillator should be used to achieve better performance from the mixers.

- The RMS noise per ground resolution element must be less than 0.8 K for improved target information, when both channels are added.
- A calibrated signal is injected into the feed horn at the end of each scan line to provide for temperature calibration.
- For optimum results, the relative accuracy of +/− 1 K and the absolute accuracy of +/− 10 K must be achieved.
- The radiometer output signals need to be digitized, converted to a brightness temperature, and recorded on a digital cartridge tape by a micro-computer to maintain the radiometer performance data.

*6.11.3 Gunn Diode Oscillators Requirements for Mixers*
*for Use in mm–Wave Radiometers*

It is critically important to mention that mixer performance is strictly dependent on the local oscillator performance specifications in terms of power capability, frequency stability, and spurious levels. Comprehensive research studies undertaken by the author on GaN-based negative differential resistance (NDR) diode oscillators using Gunn diode design criteria reveal that large-signal analysis of the GaN NDR diode oscillators is necessary. These studies indicate that THz signal generation by the GaN NDR diodes offers significantly higher frequency and RF power capability than GaAs Gunn diodes due to higher electron velocity and reduced time constants. Preliminary studies seem to indicate that microwave diodes with NDR, such as GAAS or InP Gunn diodes, are the preferred devices for generation of microwave signals at mm-wave frequency and higher output power levels in excess of 200 mW at frequencies greater than 95 GHz.

Research studies performed by the author on electronic properties of III-V nitrides seem to indicate that these wide-band gap materials also exhibit bulk NDR effect with threshold field levels in excess of 80,000 V/cm. Furthermore, Monte Carlo studies of electron transport indicate that the energy relaxation time in GaN is much shorter than that in conventional III-V semiconductors. From these statements, one can conclude that short energy relaxation time and high threshold field exceeding 80,000 V/cm are necessary to generate higher output power at higher mm-wave frequencies.

## 6.12  mm-Wave Forward-Looking Imaging Radiometers Using GaN and InP Diodes

These mm-wave sensors are referred to as mm-wave forward-looking infrared radiometers (FLIRs) and can be used for numerous commercial and military applications where high resolution and rapid image generation are the principal requirements. This particular sensor is best suited for tactical missions. The latest FLIRs are operating near the earth surface, in high altitude aircrafts, and in low and synchronous orbit satellites. In general, major applications include surface sensing of geographic features, weather prediction, and object location. Future applications will require high resolution and rapid generation of high-quality image generation for military, scientific, and commercial applications. Scientists predict that the new generation of these sensors will deploy the latest GaN-based NDR diode oscillators for high power capabilities at frequencies exceeding 200 GHz. The applications of these imaging sensors are summarized in Table 6.6 [5].

It is critically important to point out that mm frequencies offer significant advantages such as the ability to provide high-quality images under cloudy, hazy, and foggy weather conditions. This kind of operation is of vital importance in tactical applications, battlefield environments, landing under adverse weather conditions, adverse weather aircraft landing operations, undertaking of surveillance and

**Table 6.6**  Applications of New Generation of FLIRs

| PLATFORM | IMAGE BEST SUITED FOR | REMARKS |
| --- | --- | --- |
| Satellite | Surface targets (land and sea) | Low orbit cases |
| Spacecrafts | Planets | Multi-frequency |
| Aircraft (fixed wing) | Tactical targets and precision landing | Navigation and precision weapon delivery |
| Missile drone | FEBA engagement | Guidance, reconnaissance |
| Helicopter | Battlefield, surveillance | Real-time data collection, penetration in FEBA |
| Tower | Area and perimeter security | Real-time data collection, scene |
| Ship | Horizontal scanning for ships cruise missiles and aircrafts | High spatial resolution |
| Land mobile | Battlefield surveillance | Covert missions, real-time data collection |
| Man portable | Perimeter defense | Real-time data collection |

*Remark:* FEBA stands for Forward Edge Battle Field

**Table 6.7**    Atmospheric Attenuation and Zenith Sky Temperature (K) as Function of Frequency (GHz)

| FREQUENCY, (GHZ) | ATMOSPHERIC STATUS | HORIZONTAL PATH | | ZENITH SKY TEMPERATURE (K) | | | |
|---|---|---|---|---|---|---|---|
| | | SEAL | 4 KM (HT) | ZENITH TOTAL | CLEAR | PARTLY CLOUDY | MEDIUM RAIN |
| 95 | Window | 0.4 | 0.10 | 1.0 | 60 | 150 | 240 |
| 118 | Absorption ($O_2$) | 2.0 | 0.90 | 100+ | 290 | 290 | 290 |
| 140 | Window | 1.5 | 0.15 | 1.5 | 130 | 200 | 250 |
| 183 | Absorption ($H_2O$) | 30 | 3.0 | 100+ | 290 | 290 | 290 |
| 220 | Window | 5.5 | 0.20 | 2.6 | 170 | 260 | 290 |
| 320 | Absorption ($H_2O$) | 35.0 | 6.0 | 100+ | 290 | 290 | 290 |
| 350 | Window | 9.0 | 0.80 | 6.0 | 220 | 280 | 290 |

*Remark:* Under clear air conditions, standard pressure is assumed to be 7.5 grams/m³ of $H_2O$.

reconnaissance missions, and remote sensing of critical battlefield signatures.

One must understand the atmospheric attenuation or loss and zenith sky temperature at frequencies 95 GHz and above. Typical atmospheric attenuation and zenith sky temperatures as a function of RF frequencies have been summarized in Table 6.7 [5].

It is important to mention that the radiometric temperature of a surface consists of both the thermally emitted power and the reflected power. The emissivity of a surface is dependent on the surface conditions and the ratio of surface brightness temperature to the brightness temperature of a black body at the same physical temperature, which varies as follows: 0, <ℰ, <1.

### 6.12.1 Estimation of Radiometric Temperature as a Function of Background Surfaces under Clear and Moderate Rain Conditions

For precision system design it is desirable to use the standard radiometric temperatures as a function of surface conditions under clear and moderate rain conditions. It is important to mention that the radiometric temperature of a surface consists of a thermally emitted power component (Pℰ) and a reflected power component (Pre). It should be stressed that the emissivity (ℰ) of a surface is defined as the ratio of surface brightness temperature to the brightness temperature of a blackbody at the same physical (kinetic) temperature.

**Table 6.8**  95-GHz Radiometric Temperatures as a Function of Background Surfaces under Clear and Moderate Rain Conditions

| SURFACE CONDITION | CLEAR CONDITIONS | MODERATE RAIN (K) |
|---|---|---|
| Short grass | 240 | 240 |
| Tall grass | 250 | 250 |
| Bare dirt | 220 | 230 |
| Ice | 210 | 220 |
| Concrete | 200 | 210 |
| Gravel | 200 | 210 |
| Water (still) surface | 190 | 200 |

In addition, the emissivity is strictly a function of the aspect angle and the apparent radiometric temperature of a surface also depends on the aspect angle, which is also known as grazing angle or depression angle. Typical 95-GHz radiometric temperatures as a function of background surfaces under clear and moderate rain conditions are summarized in Table 6.8 [5].

### 6.12.2 Types of Tracking Radiometers for Tactical Deployment

*6.12.2.1 Angular Tracking Function*  Comprehensive research studies seems to indicate that mm-wave tracking radiometers are widely deployed in tactical applications and low orbit satellites. Tracking radiometers are best suited for target designation, weapon delivery, and terminal guidance functions. Angle measuring systems such as conical scan, sequential lobbing, or mono-pulse are implemented to generate an error signal indicating magnitude and phase parameters, which will indicate the angle error from the tracking radiometer-to-target. Note that radiometric integration time is strictly a function of situation dependence, target extent, tracking accuracy, and tactical radiometric geometrical parameters as shown in Figure 6.4.

*6.12.2.2 Target Detection Function*  This radiometer is a very desirable sensor for target detection. The detection is strictly dependent on the detection probability based on a particular situation and on background surface conditions. Note the nature of background, which could be uniform or cluttered, the background dynamics such as sea state and wave level, and the physical size of the target. But

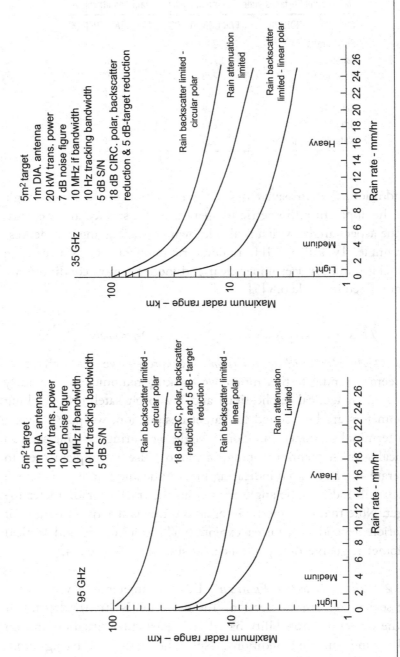

**Figure 6.4** Radiometric integration time.

the emissivity difference between target and background is of critical importance, and must be the taken into account. It must be stressed that the difference could be as high as 0.9 for an armored vehicle and as low as 0.15 for an airfield on a grassy field. In addition, the impact of weather conditions must be considered.

It should be remembered that the probability of detecting unresolved targets could be as high as 0.9 with signal-to-noise ratio of 4 dB. Furthermore, if the target is too large to be contained in the antenna beam, the signal-to-noise should be increased by the squire root of n, where n is number of separate beam positions to cover the target completely within the antenna beam.

There are other applications of mm-wave radiometers but the author has chosen not to discuss them further to limit the scope and size of this text.

## 6.13 Summary

This chapter is dedicated to the deployment of rare earth materials in the design and development of mm-wave radar transmitters for radars, mm-wave radiometers, and high-power EO lasers. Research studies undertaken by the author indicate that the deployment of RE elements, oxides, compounds, and alloys offer unique advantages in overall performance such as efficiency, reliability, and reduction in weight, size, and power consumption. The author has identified that, when RE oxides and alloys are inserted in the design of components and sub-systems, this brings significant advantages and applications in various commercial, industrial, and research fields.

SmCo magnets are currently being used in the design and development of electric motors and generators, which are the most critical components of all-electric and hybrid electric automobiles. The implementation of these rare earth magnets in commercial vehicles has demonstrated higher efficiency, lower temperatures inside the vehicles, and significant improvement in reduction of weight, size, and cost of certain components.

The use of SmCo permanent magnets in the design of radar transmitters and TWTAs has resulted in a significant improvement in RF power output, efficiency, thermal performance, and reliability under severe thermal and mechanical environments. In addition, a reduction in the weight and size of the radar transmitter has been observed, which is considered most desirable in airborne applications. It should be stated that the use of SmCo PPMs in TWTAs has not only improved the RF power output, but also reduced the thermal dissipation. In addition, these PPM magnets focus the magnetic field over the entire length of the helix, which produces uniform gain of the tube, reduces spurious responses, and keeps excellent AM-to-PM performance under high power conditions. Particularly, during the after-burner operation, the TWTA keeps the jamming power constant during the entire jamming duration, which is considered most remarkable in enemy jamming situations.

Solid state microwave sources in the mm-wave spectrum are best suited for mm-wave radiometers. The author has identified mm-wave microwave diode sources which generate adequate power at upper mm-wave frequencies. Mm-wave radiometers and IR lasers can use the IR energy produced by solid state microwave diodes such as GaAs, GaN, HEMT, p-HEMT, Gunn IMPATT, INP and GaN-based NDR diodes. Preliminary calculations made by the author indicate that GaN-based NDR diodes are most efficient in generating power levels exceeding 200 mW in upper microwave regions. These microwave diodes using rare earth elements have demonstrated unique RF power levels, which are best suited for radiometers in tactical applications. Rough calculations seem to indicate that GaN NDR diodes produce mm-wave power levels at 95 GHz to 350 GHz for radiometric applications. Such radiometers are best suited for detecting and tracking ground and airborne targets under clear and moderate rain conditions. These radiometers can be installed on helicopters, tactical aircrafts, and low-orbit satellites for the detection and tracking of airborne targets.

Solid state microwave devices such as HEMTs, p-HEMTS, and GaAs are best suited for RF MMIC amplifiers operating at frequencies close to 95 GHz. These devices deploy rare earth elements and offer excellent RF amplifier characteristics with minimum weight and size; therefore they are best suited for airborne and space applications. It is critically important to mention that certain solid state amplifiers could be designed and developed to meet radiation hardening specifications. These solid state MMIC amplifiers are best suited for airborne and satellite applications, where minimum size, light weight, and low power consumption are the basic requirements.

There are specific rare earth compounds and oxides which play critical roles in some applications. Some rare earth oxides are best suited for fuel cell applications because of their high melting temperatures and excellent compressive strength. Zr oxide ($Zr_2O_2$) is widely deployed in the design and development of high-temperature fuel cells due to the excellent thermal properties of Zr. Dy is widely used for fuel control rods in nuclear power reactors. Er is widely used in the design of fiber optic amplifiers, which plays an outstanding role in compensating for the insertion loss in very long fiber optic communication lines. La oxide plays a critical role in the development of superconducting thin films for microwave devices. Li is best suited in the development of batteries for cell phones, computers, electronic toys, electronic savers, automobiles, aircraft starters, and a host of mechanical tools.

Y oxide is widely deployed in the manufacturing of YIG filters, which are used in the design and development of microwave filters where minimum insertion loss, high rejection, and sharp cutoff capability are the main requirements. In addition, YIG is considered the most ideal material for the development of transducers for the electronic industry. It is important to mention that this material offers lower transducer dimensions and the optimum impedance match needed for low-input VSWR and minimum insertion loss.

Thin films of rare earth compounds such as YBCO, Thallium Barium Calcium Copper oxide (TBCCO) and other compounds play important roles in the design and development of superconducting thin films, which are widely used for microwave RF components such as filters with sharp cutoff frequencies and minimum insertion loss. Some of these superconducting thin films are used in the development of superconducting substrates for deposition of thin films. It is critically important to stress that superconducting thin films can be deposited on superconducting substrates only if optimum performance of the microwave component is desired.

Laser-doped crystals using rare earth elements such as Ho, Eu, Y, YLF, $T_m$, Y, and YAG are best suited for the design and development of IR lasers emitting at IR wavelengths. The most popular IR laser source configurations are shown in Figure 6.1 and Figure 6.2.

Rare earth oxides such as $Zr_2O_2$ and $Y_2O_2$ play critical roles in the design and development of high-power fuel cells. Fuel cells scientists feel that the thermal coefficient of expansion of these rare earth oxides is of critical importance. These rare earth oxides yield optimum fuel cell efficiency, the highest possible output power, and utmost safety. Research studies performed by the author indicate that a PEM offers optimum performance in terms of power, efficiency, and safe operation of the device. A fuel cell configuration as shown in Figure 6.3 deploys $H_2O_2$ and provides a most compact device with minimum cost and complexity. This fuel cell design offers minimum device dimensions and utmost safety.

Rare earth–based GaAsP/InP diodes are best suited for QW, DQW, and MQW laser configuration. Note that the QW diode offers higher optical power with improved differential quantum conversion efficiency. A DQW diode offers higher optical power than a QW diode, and an MQW diode offers optical power outputs exceeding those of QW and DQW diodes. Preliminary studies undertaken by the author indicate that InP laser diodes offer excellent lifetime characteristics. Studies further indicate that the InGaP heterostructure diodes generate IR energy over 1.55 to 2.10 spectral region without cryogenic cooling, thereby providing IR lasers with minimum cost and complexity. Comprehensive research studies indicate that the rare earth–based InGaAs-DQW diode can deliver optical power output in the order of several hundred mW per facet. Computer analysis performed on these diodes reveals that total DQE can be as high as 47% at room temperature. Furthermore, the analysis indicates that the threshold current decreases from 46 mA at room temperature to 14 mA at cryogenic temperature at $-10°C$. Note that these diodes offer FWHM laser beam width in the perpendicular plane, which remains constant irrespective of the power output. These quantum well diodes can be installed in the ridge waveguide of small aspect ratio, thereby realizing substantial reductions in the weight, size, and cost of the laser package. Tunable laser requires a tetravalent chromium laser crystal $Cr^4:Ca_2 GeO_4$ diode, which offers tuning range from 1.3 to 1.6 micro range. Such lasers are best suited for applications in heavily defended electronic environments.

MM-wave radiometers are best suited for surveillance and reconnaissance of ground and airborne targets. It is interesting to mention that 95 GHz is the most popular frequency for radiometers as well as for airborne radars. As far as detectors are concerned, rare earth elements such as In, Cd, and telluride are best suited for radiometers. These detectors offer high detection capability with high probability, which is essential for high-resolution target images.

Radar transmitters, EIOs, travelling wave tubes, and gyrotrans are available to meet power requirements. For low to medium transmitter power requirement, solid state power sources such as Gunn diodes, IMPATT diode oscillators, and GaN NDR diode oscillators are considered most ideal. Note that these solid state diode RF sources offer minimum weight, size, and cost.

It is important to mention that InP and GaN NDR diode oscillators can be deployed to drive mixers where high LO power is required to optimize the mixer performance. As mentioned earlier, GaN NDR diodes offer high power output at mm-wave frequencies exceeding 150 GHz. Note that mm-wave FLIR radiometers can use GaN NDR diode oscillators to design an extremely light mm-wave transmitter package for airborne applications. Note that mm-wave sources offer high-resolution images and rapid-generation high-quality target images for tactical and scientific applications with minimum cost.

A high-resolution radiometer requires mm-wave transmitters operating at 95 GHz or higher. The radiometer can be installed on a helicopter, drone, aircraft, or tower. To meet high-resolution requirements, a radiometer transmitter requires design frequency greater than 95 GHz, antenna beam close to 1 degree, pulse better than 0.5 microseconds, and antenna side lobe levels better than 35 dB (RMS). These are preliminary radiometer antenna performance parameters to provide high-resolution images.

# References

1. A.R. Jha, *Rare Earth Materials: Properties and Applications*, CRC Press, Taylor & Francis Group, Boca Raton, Florida, 2014 Edition, pp. 59–61.
2. A.R. Jha, *Rare Earth Materials: Properties and Applications*, CRC Press, Taylor & Francis Group, Boca Raton, Florida, 2014 Edition, pp. 312–314.

3. A.R. Jha, *Next Generation of Batteries and Fuel Cells for Commercial, Military and Space Applications*, CRC Press, Taylor & Francis Group, Boca Raton, Florida, 2012 Edition, pp. 312–315.

4. A.R. Jha, *Rare Earth Materials: Properties and Applications*, CRC Press, Taylor & Francis Group, Boca Raton, Florida, 2014 Edition, pp. 70–71.

5. J. M. Schuchardt et al., "The Coming of mm-Wave FLIRs", Georgia Institute of technology, Engineering Experimental Station, Alabama (GA), published in *Microwave Journal*, June 1981, pp. 45–52.

# Index

Printed in the United States
by Baker & Taylor Publisher Services